Lecture Notes in Networks and Systems

Volume 108

The series "Lecture Notes in Networks and Systems" publishes the latest developments in Networks and Systems—quickly, informally and with high quality. Original research reported in proceedings and post-proceedings represents the core of LNNS.

Volumes published in LNNS embrace all aspects and subfields of, as well as new challenges in, Networks and Systems.

The series contains proceedings and edited volumes in systems and networks, spanning the areas of Cyber-Physical Systems, Autonomous Systems, Sensor Networks, Control Systems, Energy Systems, Automotive Systems, Biological Systems, Vehicular Networking and Connected Vehicles, Aerospace Systems, Automation, Manufacturing, Smart Grids, Nonlinear Systems, Power Systems, Robotics, Social Systems, Economic Systems and other. Of particular value to both the contributors and the readership are the short publication timeframe and the world-wide distribution and exposure which enable both a wide and rapid dissemination of research output.

The series covers the theory, applications, and perspectives on the state of the art and future developments relevant to systems and networks, decision making, control, complex processes and related areas, as embedded in the fields of interdisciplinary and applied sciences, engineering, computer science, physics, economics, social, and life sciences, as well as the paradigms and methodologies behind them.

**** Indexing: The books of this series are submitted to ISI Proceedings, SCOPUS, Google Scholar and Springerlink ****

More information about this series at http://www.springer.com/series/15179

Douglas Gorham · Nnamdi Nwulu

Engineering Education through Social Innovation

The Contribution of Professional Societies

Douglas Gorham
IEEE Educational Activities Department
Kissimmee, FL, USA

Nnamdi Nwulu
Department of Electrical and Electronic Engineering Science
University of Johannesburg
Johannesburg, South Africa

ISSN 2367-3370 ISSN 2367-3389 (electronic)
Lecture Notes in Networks and Systems
ISBN 978-3-030-39008-2 ISBN 978-3-030-39006-8 (eBook)
https://doi.org/10.1007/978-3-030-39006-8

This Springer imprint is published by the registered company Springer Nature Switzerland AG
The registered company address is: Gewerbestrasse 11, 6330 Cham, Switzerland

Preface

Abstract Educating engineers in the twenty-first century poses challenges not faced in any other centennial. Professional societies are in a unique position to assist and support academia, industry, governmental bodies and non-governmental entities in addressing the needs of engineering education through programmes and activities leveraging social innovation. These programmes are focused on making an impact on pre-university education teachers and their students, university faculties and students, programme quality and the public.

The following chapters summarise the context of engineering education in the twenty-first century and describe how prominent professional societies, such as the Institute of Electrical and Electronics Engineers, the South African Institute of Electrical Engineers, the Institution of Engineering and Technology, the China Association for Science and Technology and the Society of Automotive Engineers International, among others, are supporting and promoting engineering education at the pre-university level, tertiary level and the informal sector.

Kissimmee, USA — Douglas Gorham
Johannesburg, South Africa — Nnamdi Nwulu

Preface

Contents

Chapter 1
The Nexus Between Engineering Societies and Engineering Education

Abstract Engineering as a profession and engineers celebrated significant accomplishments in the 20th century that enhanced and transformed every facet of human and social life. The 21st century poses challenges as daunting as any from past centuries and thus adds to the changing curricular landscape for engineering educators and institutions. Today's engineering schools are facing unique obstacles coupled with high expectations to produce graduates who can function and thrive in the 21st century. Technological advancements and societal changes in the 21st century place a demand on engineers to design, create and deploy technologies and solutions that satisfy complex technical, ethical and sustainability requirements.

1.1 Engineering Societies: History, Context and Purpose

Professional engineering societies began forming in the 19th century and early 20th century, focusing on the engineering profession and engineers. Some of the early engineering societies include the Institution of Engineers of Ireland, founded in 1835 (Institution of Engineers of Ireland 2018); Verein Deutscher Ingenieure (Association of German Engineers), founded in 1856 (Verein Deutscher Ingenieure 2018); American Society of Mechanical Engineers (US), founded in 1880 (American Society of Mechanical Engineers 2018); The American Society of Civil Engineers (US), founded in 1852 (American Society of Civil Engineers 2018); Institute of Electrical and Electronics Engineers (IEEE—US), founded in 1884 (Institute of Electrical and Electronics Engineers, Inc. 2018); Japan Society of Civil Engineers, founded in 1914 (Japan Society of Civil Engineers 2018); Institution of Chemical Engineers (United Kingdom—UK), founded in 1922 (Institution of Chemical Engineers 2018); and The Institution of Engineers Australia, incorporated in 1926 (Institution of Engineers Australia 2018).

Not unlike today, the organisers of these early professional engineering organisations focused on improving standards, promoting research and sharing knowledge through publications and conferences.

Today professional engineering societies continue to embrace and expand the areas of focus developed by the early organisers. Many professional societies dedicate

D. Gorham and N. Nwulu, *Engineering Education through Social Innovation*, Lecture Notes in Networks and Systems 108, https://doi.org/10.1007/978-3-030-39006-8_1

Member-benefit professional associations
- Exist primarily to create value for their members
- Offer various products and services to members
- Networking and events
- Advocacy on behalf of the members
- Do not have a statutory mandate and obligation to promote and protect the public interest

Designation-granting associations
- Similar to associations, but in addition offer one or more designations
- Designations are seen as an added-value product or service offered to members
- Certification is not a requirement of membership, but membership is required to maintain right to use designation

Certifying bodies
- Focus is on the designation(s) as a product
- Few member services
- One must be certified to maintain membership in the certifying body
- Do not have a statutory mandate and obligation to promote and protect the public interest

Professional regulatory bodies
- Exist primarily to promote and protect the public interest and not to serve the interests of the professionals under regulation
- Exercise authorities delegated by law pursuant to statute
- Establish, maintain, develop, and enforce standards of qualification, practice, and conduct

Fig. 1.1 Four broad classes of professional associations (Balthazard 2017)

resources to social responsibility programmes and activities to promote engineers and engineering.

There are essentially four kinds of professional societies: member benefit professional associations, associations that offer designations, certifying bodies, and professional regulatory bodies (Balthazard 2017). The four types of associations and their characteristics are described in Fig. 1.1. Depending on their focus, they contribute significantly to society. For the purpose of this book, only member benefit professional associations will be considered.

1.2 Context of Engineering Education

The engineering profession celebrated significant accomplishments in the 20th century (US National Academy of Engineering 2008). These accomplishments range from the generation of electrical power to the advent of the internet and numerous advances in medicine and health. These examples are a sampling from the 20th century where engineers and engineering undoubtedly made significant contributions to humanity.

At its core, engineering is the application and integration of mathematics, science, technology, economics, and social and practical knowledge to improve the material living standards of humans. The UK Royal Academy of Engineering (RAE) 2007 report, "Educating Engineers for the 21st Century" (Royal Academy of Engineering 2007), describes the connection between engineers and a nation's economy:

> No factor is more critical in underpinning the continuing health and vitality of any national economy than a strong supply of graduate engineers equipped with the understanding, attitudes and abilities necessary to apply their skills in business and other environments.

As such, universities are tasked with the complex challenge of delivering pedagogically effective, high-quality education resulting in engineering graduates who possess technical competence, global sensitivity, a team approach, creativity and innovation, strong interpersonal skills, business expertise, social awareness, ethics and ethical behaviour, and the ability to work in multiple disciplines and environments.

1.3 Challenges Facing Engineering Education

1.3.1 Changing Engineering Landscape

The advent of the Fourth Industrial Revolution (4IR), which has birthed a variety of disruptive concepts and technologies, will also drastically change the education landscape in general and engineering education in particular. A report by the Global Schools Leadership Alliance (GLA 2016) posits that the current state of education does not adequately prepare students for the future and that "disruptive innovation awareness" is a key attribute that must be inculcated in education.

This has led to strident calls for a paradigm shift in the way engineering education is conducted. It has been advocated that the science-based engineering curriculum, which has been used since the 20th century and is suitable for "complicated" engineering systems, should be replaced by an engineering curriculum method that prepares students for "complex" engineering systems (Kastenberg et al. 2006). The distinction between "complicated" and "complex" engineering systems lies in the fact that complicated systems can be understood by comprehending the behaviour of their constituent sub-systems, while complex systems exhibit different behaviour as a whole, can only be understood as a whole and cannot be deciphered by a comprehension of their constituent sub-systems (Kastenberg et al. 2006).

1.3.2 Challenges with Pre-University STEM Education

To understand the challenges facing engineering education properly, one must consider it from the more extensive perspective of the challenges bedevilling education in the science, technology, engineering and mathematics (STEM) fields, as engineering education does not exist in a vacuum. In spite of the huge amounts of money poured into STEM education, outcomes of STEM programmes in most nations of the world leave much to be desired. A key conclusion reached is that early interventions, especially at the pre-university level, are key. Additional challenges include the perception of engineering and a shortage of qualified pre-university teachers in key areas. RAE's 2007 report, "Educating Engineers for the 21st Century" (Royal Academy of Engineering 2007) highlights these issues:

- In the secondary schools, where students make decisions about the university courses they will pursue, there is an acknowledged shortage of teachers in mathematics and physics, the essential precursors of undergraduate engineering studies.
- To fill the pipeline, more must be done to ensure that school students, parents and teachers perceive engineering as an exciting and worthwhile subject that offers stimulating and well-paid careers.

1.3.3 Lack of Public Awareness of Engineering at the Pre-University Level

There is a need for engineering educators and engineering schools to develop comprehensive promotional programmes to increase the awareness of engineers and engineering among pre-university teachers, their students, government and non-governmental entities, and the public.

Significant and important contributions have been made by engineers across the world in solving global problems and increasing the health and wealth of countries. However, the nature of engineering, its scope, diversity and impact on society are all too often poorly understood by pre-university teachers, their students, government entities, non-governmental organisations (NGOs) and the public. The engineering community is often not engaged in public policy discussions and process. This issue is about influencing the perception of key groups, including pre-university teachers, their students, NGOs, appropriate governmental organisations and the public.

Contributing to the lack of understanding of what engineers do is the broad spectrum of engineering disciplines and branches. This has the effect of creating a barrier that blocks key groups and organisations from a better understanding and awareness of what engineers do and the significant contributions engineering makes to society.

To combat this trend, additional programmes and resources need to be deployed to spur on and capture the imagination of young people to pursue engineering careers. This needs to be achieved on a global scale. In addition, young people must be enlightened about the societal, economic and environmental impact engineering and by extension engineers have on the world. Conveying an engineering message can feature several strategies, including informal education (e.g., science museums), pre-university schools (including teachers, counsellors and students), a robust and effective online presence and programmes for university faculty and students. A key component of all promotional efforts is a focus on recruiting under-represented groups, including women and people from poor socioeconomic backgrounds.

1.4 Rethinking Engineering Education Delivery

Multiple reports and studies identify the need to change how engineering education is delivered. Reports including “Thinking Like an Engineer: Implications for the

Education System" (RAE 2014), "Engineering: Issues, Challenges and Opportunities for Development" (UNESCO 2010), "The Engineer of 2020: Visions of Engineering in the New Century" (NAE 2004) and: "Educating the Engineer of 2020" (NAE 2005) all stress the need to change the curricula and pedagogical approach of engineering education.

One of the recommendations included in the US NAE Report; "The Engineer of 2020: Visions of Engineering in the New Century" (National Academy of Engineering 2004) highlights this need:

> As well as delivering content, engineering schools must teach engineering students how to learn, and must play a continuing role along with professional organizations in facilitating lifelong learning ... (p. 55)

The RAE's 2007 report, "Educating Engineers for the 21st Century" (Royal Academy of Engineering 2007) echoes the need for technical knowledge and the ability to apply it in a real-world setting. The report states: "Engineering courses must develop in line with the real and constantly evolving requirements of industry." The report further emphasises the need for using effective pedagogical strategies, based on industry needs, to deliver the curricula and support faculty in attaining new knowledge and skills. The report states:

- ... academic staff need the time and resources to implement new approaches to engineering learning and teaching.
- There is a need to embed multidisciplinary approaches based on systems thinking, with strong industry links, within all engineering courses.

Engineering educators and schools of engineering can benefit from the 2014 report of the Royal Academy of Engineering: "Thinking like an Engineer: Implications for the Education System" (Royal Academy of Engineering 2014). This report highlights how engineers think and describes six engineering habits of mind (EHoM) that can inform and assist in reshaping how engineering is taught at the pre-university and university levels. Figure 1.2 provides a description of EHoM.

Utilizing the engineering design process as a pedagogical principle, EHoM can be incorporated into this process. Even though this process is fraught with inherent implementation difficulties, it can be adapted to all education strata. In Fig. 1.3, the engineering design process is detailed.

The engineering design process depicted above suggests a step-by-step approach. This is not as straightforward in practice. Multiple revisions or iterations play a significant role in each step. It is likely that within each of the stages shown in Fig. 1.3, there might be various iterations before moving on to the next stage. The engineering design process has been described by Zeid et al. (2013) as "systematically organised chaos where every step has more than one solution and more than one method."

Table 1.1 summarises the issues of the 21st century, the needed attributes of effective engineers in the 21st century, and the desired characteristics of engineering education in the 21st century. Reports and studies focused on the needs and challenges in engineering education provide the basis for Table 1.1. These include (Royal

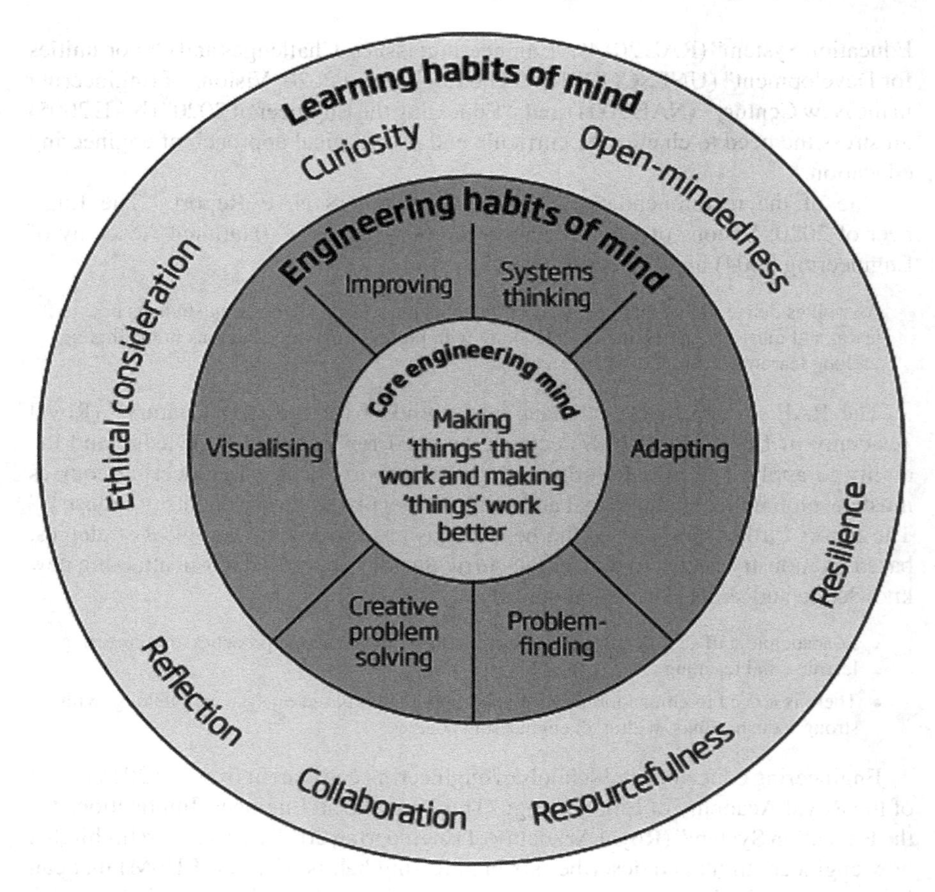

Fig. 1.2 Engineering habits of mind (Royal Academy of Engineering 2014)

Academy of Engineering 2006, 2007, The World Bank 2014, United Nations Educational, Scientific and Cultural Organization 2010, Duderstadt JJ 2008, National Academy of Engineering 2004, 2005, International Engineering Alliance 2016).

1.5 Industry Perspective

It is crucial that engineering educators and institutions consider the needs and perspectives of industry. The RAE published a report in 2006 entitled "Educating Engineers for the 21st Century: The Industry View." (Royal Academy of Engineering 2006). This report highlights future needs of engineers and engineering:

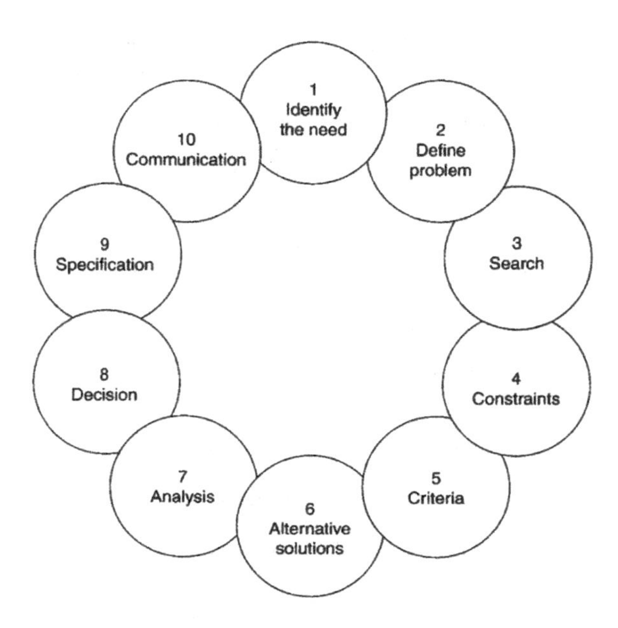

Fig. 1.3 The engineering design process (Eide 2002)

Table 1.1 Engineering education in the 21st century

Global issues of the 21st century	Attributes of effective engineers	Desired characteristics of engineering education
Knowledge economy	Analytical skills	Centre of study
Globalisation	Practical ingenuity	Training based on discovery and constructivism
Changing demographics	Creativity	The perspective of the systems
Technological change	Communication	Learn how to memorise
Technological innovation	Leadership	Request scientific methods of analysis
Global sustainability	Teamwork	Issue-tackling focused on a squad
Energy	Professionalism	Making engineers ready for a worldwide economy
Global poverty and health	Dynamic, agility, resilience and flexible	Creating a research foundation
Infrastructure	Lifelong learners	Building engineering without limits
Urbanisation	Function in a global economy	
	Principles of business and management	

> The research finds that the pace of change in the industry is expected to intensify in both the technological and nontechnological domains. Themes that emerged include an increased need for firms to focus on solving customer problems; a growing requirement to provide system solutions to those problems; and the increasing complexity of the management task. This management complexity is paralleled by growing technological complexity and interdependence at all levels. Another important factor is globalisation which will continue to affect both the demand and the supply-side of the industry. In particular, the rapid growth in placing work offshore poses a real challenge that will force UK engineering in all areas to concentrate on higher value-added activities. As a result, there will be a premium on innovation and creativity to respond to the challenges of this turbulent environment.

The RAE 2006 report goes on to describe the engineering graduate of the future:

> The research synthesises these findings into a picture of the graduate engineer of the future. At the heart lie the defining and enabling skills that form the core competencies of the engineering graduate. Whilst other professions will have special skill sets, it is this particular combination of skills that marks out the engineering graduate and underpins the roles that industry will need such graduates to undertake. Three such roles are identified. Firstly, the role of engineer as specialist recognises the continued need for engineering graduates who are technical experts of world-class standing. Secondly, the engineer as integrator reflects the need for graduates who can operate and manage across boundaries, be they technical or organisational, in a complex business environment. Thirdly, the engineer as change agent highlights the critical role engineering graduates must play in providing the creativity, innovation, and leadership needed to guide the industry to a successful future. This is a vision of the future that underlines the vital importance of undergraduate engineering education to the UK engineering industry. At the same time, however, it emphasises the reciprocal responsibility of the engineering industry in ensuring the future excellence of undergraduate engineering education in the United Kingdom.

Industries such as IBM and Siemens and the US National Science Foundation define the need for "T-shaped" engineers, first described by David Guest in 1991 (Guest 1991), as "those with deep knowledge and expertise in their discipline and a broad breadth of cross disciplinary knowledge and boundary crossing capabilities including an understanding of business context and human as well as social aspects of engineering, communication, systems perspective, lifelong learning, ability to innovate and the ability to adapt to a changing environment." (Fig. 1.4).

The vertical bar of the 'T' is representative of the expertise of the engineer in a single field, while the horizontal bar depicts the skills and collaborative expertise in disciplines other than one's own.

1.6 Structure of the Book

Educating engineers in the 21st century brings about challenges not faced in any other centennial. Professional societies are in a unique position to assist and support academia, industry, governmental bodies and non-governmental entities in addressing the needs of engineering education and the perception of engineers and engineering through programmes and activities leveraging social innovation. This book seeks to describe how several professional societies are supporting and promoting

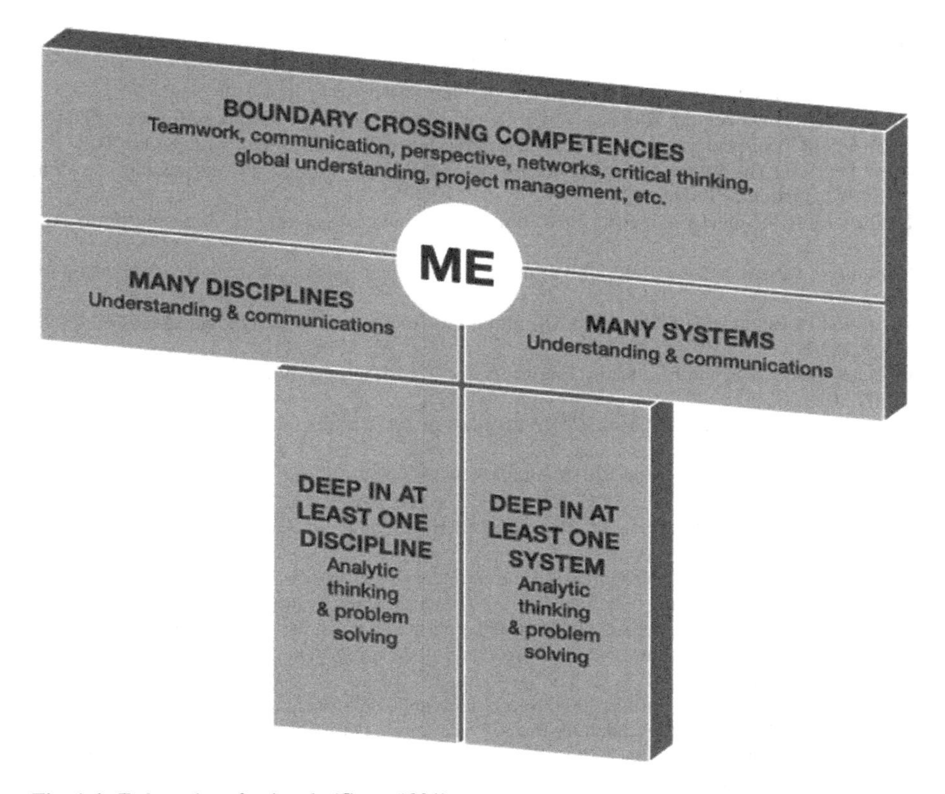

Fig. 1.4 T-shaped professionals (Guest 1991)

engineering education in the areas of pre-university education, university education and informal education through social innovation. Chapter 2 considers the support professional bodies give to pre-university teachers and their students. Chapters 3 and 4 delve into respectively direct and indirect engagements professional societies use to support engineering education. Chapter 5 focuses on professional societies supporting activities in the informal education sector. Chapter 6 focuses on how professional societies can establish and/or expand their efforts to support engineering education and serves as the concluding chapter.

References

American Society of Civil Engineers (2018) https://www.asce.org. Accessed on 28 July 2018

American Society of Mechanical Engineers (2018) https://www.asme.org. Accessed on 15 July 2018

Balthazard C (2017) The four types of professional organizations. Retrieved from https://www.linkedin.com/pulse/four-types-professional-organizations-claude, 10 Oct 2018

Duderstadt JJ (2008) Engineering for a changing world: a roadmap to the future of American engineering practice, research, and education. University of Michigan
Eide A (2002) Engineering fundamentals and problem solving. McGraw-Hill, New York
Global Schools Leadership Alliance (2016) Retrieved from http://www.schoolsalliance.org/news/2016/2/3/global-perceptions-of-education-and-the-fourth-industrial-revolution, 8 Oct 2018
Guest D (1991) The hunt is on for the Renaissance Man of computing. The Independent. London, United Kingdom, 17 Sept 1991
Institution of Chemical Engineers (2018) http://www.icheme.org/about_us. Accessed on 19 Apr 2018
Institution of Engineers Australia (2018) https://www.engineersaustralia.org.au/. Accessed on 19 Apr 2018
Institution of Engineers of Ireland (2018) http://www.engineersireland.ie/home. Accessed on 18 Apr 2018
Institute of Electrical and Electronics Engineers, Inc. (2018) https://www.ieee.org. Accessed on 19 July 2018
International Engineering Alliance (2016) 25 years of the Washington Accord. Wellington, New Zealand
Japan Society of Civil Engineers (2018) http://www.jsce-int.org/about. Accessed on 20 Apr 2018
Kastenberg W, Hauser-Kastenberg G, Norris D (2006) An approach to undergraduate engineering education for the 21st century. In: 36th ASEE/IEEE Frontiers in Education Conference, 28–31 Oct 2006, San Diego, CA
National Academy of Engineering (2004) The engineer of 2020: visions of engineering in the new century. National Academy of Sciences: The National Academies Press, Washington, D.C., USA
National Academy of Engineering (2005) Educating the engineer of 2020: adapting engineering education to the new century. National Academy of Sciences: The National Academies Press, Washington, D.C., USA
National Academy of Engineering (2008) NAE grand challenges for engineering. National Academy of Sciences: The National Academies Press, Washington, D.C., USA
Royal Academy of Engineering (2006) Educating engineers for the 21st century: the industry view. London, United Kingdom
Royal Academy of Engineering (2007) Educating engineers for the 21st century. London, United Kingdom
Royal Academy of Engineering (2014) Thinking like an engineer: implications for the education system. London, United Kingdom
The World Bank (2014) Improving the quality of engineering education and training in Africa. Washington, D.C., USA
United Nations Educational, Scientific and Cultural Organization (2010) UNESCO report: engineering: issues challenges and opportunities for development. Paris, France
Verein Deutscher Ingenieure (2018) The Association of German Engineers. http://www.vdi.eu/. Accessed on 20 Apr 2018
Zeid I, Chin J, Kamarthi S, Duggan C (2013) New approach to effective teaching of STEM courses in high schools. Int J Eng Educ 291:154–169

Chapter 2
How Professional Societies Support Engineering Education: Pre-University Teachers and Their Students

Abstract Multiple reports and studies highlight the need for the global engineering community and policymakers to develop programmes to address the public's perception of engineering, to influence pre-university students and their teachers and to position engineering education as a more attractive programme of study. Many engineering societies have developed programmes and activities that support and promote engineering as a programme of study and career choice among pre-university teachers, their students, and the public. Several examples of how engineering societies are delivering programmes and activities in this area are discussed.

2.1 Pre-University Engineering Education

Pre-university engineering education is increasingly receiving attention in most nations of the world. Hitherto in the pre-university education sector, there was a great fixation on science education, which over time expanded to STEM education (Sjaastad 2010). However, engineering education had often been ignored, as science, technology and mathematics education was often researched and funded by educational institutions and professionals (de Vries et al. 2016). The increasing focus on engineering education at pre-university level stems from increasing awareness of the important role engineering and engineers play in society and the need to nurture and foster disciplinary interest among pre-university students actively (Purzer et al. 2014).

The United Nations Educational, Scientific and Cultural Organisation (UNESCO) in 2010 published "Engineering Issues, Challenges, and Opportunities for Development" (United Nations Educational, Scientific and Cultural Organisation 2010). This was UNESCO's first report on engineering, and its first attempt to focus on engineering at an international level. The report highlights the need for the engineering community and policymakers to develop programmes to address the public's perception of engineering, to influence pre-university students and their teachers and to position engineering education as a more attractive programme of study.

D. Gorham and N. Nwulu, *Engineering Education through Social Innovation*, Lecture Notes in Networks and Systems 108, https://doi.org/10.1007/978-3-030-39006-8_2

Furthermore, the report (United Nations Educational, Scientific and Cultural Organisation 2010) emphasised the need to:

- Build open approach awareness and understanding of engineering policies and regulations, affirming the role of engineering as driver of technology, social and economic development;
- Improve technical information on highlighting the pressing requirement for much better insights and markers on designing;
- Transform the engineering educational curriculum, module, and teaching methods to strategies of significance and engineering problem-solving approaches; and
- Innovate and apply science and technology more efficiently to worldwide problems and threats such as poverty reduction, sustainable development and climate change—and build greener and lower-carbon technologies as a matter of urgency.

One causal factor for the increasing attention pre-university education is garnering in engineering education is the desire to understand the key factors that spur on students to pursue an engineering career. Furthermore, pre-university educators are increasingly using engineering and its design process as a teaching aid to reinforce science, technology and mathematics concepts.

In light of this, many engineering societies have developed programmes and activities that support and seek to increase the number of entrants into engineering programmes among pre-university teachers, their students, and the public.

Several examples of how engineering societies are delivering programmes and activities in this area follow.

2.2 The South African Institute of Electrical Engineers—AfrikaBot and Bergville Community Builders

The South African Institute of Electrical Engineers (SAIEE) is focused on the future and is conscious of the fact that in order to have competent electrical engineers in the future, today's learners must imbibe the right knowledge and skills. Supporting and promoting programmes and activities focusing on pre-university students is in line with two of SAIEE's objectives: "To promote and advance education and training in electrical engineering and associated sciences in Southern Africa;" and "To increase appreciation of the role of electrical engineering practitioners and electrical engineering and associated sciences through interactions between the Institute and the Southern African community" (South African Institute of Electrical Engineers 2018).

The SAIEE has set out to increase access to and the understanding of STEM subjects for teenage students from disadvantaged communities who do not have the resources to do so themselves. These subjects are crucial to success in the engineering discipline. The SAIEE's endeavours support two programmes: AfrikaBot and the Bergville Community Builders (BCB).

2.2.1 AfrikaBot

SAIEE and the University of Johannesburg (UJ) are partnering to provide exposure to more South African teenagers to engineering concepts through involvement in AfrikaBOT, positioned as '*the world's most affordable robotics competition.*'

STEM training in developing countries is a major challenge for pre-university students. The materials, supplies and equipment needed to educate students are expensive and complex. In addition, students from disadvantaged areas are unlikely to be exposed to engineers or engineering concepts, thus limiting their options for pursuing STEM disciplines as programmes of study at university level.

As Nel et al. (2016) note:

> "The goal of establishing AfrikaBot is to prepare high school learners to study engineering at the University. Training and exposing teenagers from disadvantaged communities with no prior experience in STEM to participate in a challenge to build and programme a robot helps to equip participating teenagers with technology and entrepreneurial skills in a repressed economy. AfrikaBot achieves the above with a build-it-yourself robot that can be used after the competition to invent systems with real-world applications. It is anticipated that over time the AfrikaBot programme will influence the structure of other robotics challenges and attract a higher number of technical candidates from disadvantaged communities."

Through being offered fun and engaging transfer of technology and entrepreneurial skills, students can quickly build confidence to pursue STEM disciplines as a study or career choice programme. Participants are better prepared to set up new businesses, using entrepreneurial skills that lead to the creation of new jobs in their community in both the formal and informal sectors in Africa.

From 2016 to 2018 AfrikaBot attracted more than 350 participants from more than 20 schools located in predominantly disadvantaged communities (Figs. 2.1, 2.2 and 2.3).

2.2.2 Bergville Community Builders

Since 2003, BCB, a not-for-profit organisation, has focused on assisting high school age students from 34 schools in Bergville and surrounding areas, a rural location in northern KwaZulu-Natal. BCB's primary objective is to help high school age students obtain better understanding of their academic and career choices. BCB focuses on motivating, sharpening and broadening student's aspirations and dreams and by providing knowledge on tertiary careers, thus assisting learners to see a clearer path to realise their goals. The organisation provides the learners with necessary information that is relevant to their future career interests and provides them with guidance in the areas of life skills, financial aids and survival techniques at tertiary institutions. The group also provides students, beginning with Grade 9, with information on the importance of having a vision for one's future, setting goals and taking charge of one's life and thus one's future.

Fig. 2.1 AfrikaBot 2018

Fig. 2.2 The AfrikaBot robot (Nel et al. 2016)

Fig. 2.3 Students building the AfrikaBot robot

The BCB comprises inhabitants of Bergville who desire to create and foster an enabling environment for quality education in the Bergville community. This is essentially achieved by creating avenues for young students to learn about possible future career paths via various educational events in Bergville throughout the year. This programme has been highly successful and has led to similarly skilled professionals studying the Bergville model and exporting it to their own communities.

The SAIEE supports the BCB in its efforts to assist and motivate high school age students by providing greater insight into the electrical engineering discipline and the benefits of a tertiary education.

Apart from the SAIEE, the School of Electrical and Information Engineering of the University of the Witwatersrand is also a partner and has engaged in the organisation of many of BCB's yearly events. Some of these events include mentoring and advising young learners on probable career choices, introducing them to various concepts in science and technology and finally organising teacher training sessions for teachers in the Bergville community through train-the trainer-programmes aimed at enhancing their knowledge of science.

BCB provides a variety of customised events to increase the level of knowledge and skills of the participants. The career interests of the high school students involved are the focus of each event. Below is a summary of each programme type sponsored, organised and delivered by BCB with the support and participation of SAIEE and the School of Electrical and Information Engineering of the University of the Witwatersrand:

- Career Guidance Day

BCB organises yearly career guidance and information-sharing events. Working professionals from a variety of disciplines deliver presentations on how they are achieving their career goals. Presenters share their own qualifications and practical work experiences and offer advice and guidance. Presentations focus on entry requirements to tertiary institutions, choosing the right courses, the importance of working hard and achieving above-average results, their experiences at both tertiary institutions and in the workplace. The professionals further advise the learners on financing for their studies, how to apply for financing and the different types of financial assistance available. They also give the attendees clear information on issues related to their profession, including different career path options available for a given field of study.

This event includes role modelling for the students and motivating learners from poor backgrounds to take advantage of all help available in South Africa. An added feature includes exposing students to panel discussions on entrepreneurship to instil in students the notion that they do not have to work for someone, but can develop their own business ideas and start their own companies.

This event has yielded an increased number of learners enrolling at tertiary institutions and other vocational training colleges. BCB's annual Career Guidance Day is attended by no less than 1750 students each year. More than 15 organisations exhibit on Career Guidance Day and each organisation brings a team of people to interact with attendees.

- Career Awareness Day

This is an annual event targeting students in grade 9 to inform them about choosing a science, commerce or humanities stream in grade 10. This event assists learners in making informed decisions as to which stream to follow. They are helped by professionals in the process to determine how best to select a programme of study. Many high school teachers find this programme very useful, as they no longer have undecided learners when they enter grade 10. Nearly 4000 ninth grade students participate in Career Awareness Day each year.

- National Science Week

This is a national science, engineering and technology promotion week that takes place annually, where various science and engineering organisations are invited to promote science, engineering and technology careers. Organisations include Eskom, the South African Agency for Science and Technology Advancement, Ezemvelo KZN Wildlife, the South African Air Force, the South African National Defence Force, the University of the Witwatersrand, the South African Navy, the University of KwaZulu-Natal, JAAP Travels and Services, the South African Weather Service, SAIEE and many other organisations. They all come to showcase their technologies, activities and careers at this event for the entire week.

More than 4000 students take part in the activities during the five days of National Science Week each year. In addition, BCB's National Science Week attracts five to 10 companies that exhibit materials, make presentations and interact with students.

- Saturday Mathematics and Science Tutoring Classes

This is an intervention programme where mathematics and science teachers are used for Saturday classes to tutor students from all 34 high schools. This enables networking of the learners from different schools and provides teachers with the opportunity to share knowledge and information.

Nearly 700 students from all 34 high schools in the Bergville area participate each Saturday. These students are expected to share what they learn with their classmates upon returning to their respective schools. This programme has assisted tens of thousands of students since its inception and has helped learners win provincial Mathematics and Science Olympiads and reach the provincial top five positions in Mathematics and Science examinations.

- Train-the-Trainer

BCB's "Train-the-Trainer" programme focuses on improving and enhancing the pedagogical and knowledge competence of teachers teaching mathematics, a crucial academic discipline for success in the study of engineering.

The Train-the-Trainer initiative is a partnership between BCB and the University of the Witwatersrand. The focus on mathematics is designed to assist high school teachers in the rural area of Bergville to be more effective in the teaching and learning of mathematics. This programme leverages the expertise of postgraduate students from the School of Electrical Engineering at Wits University. More than 100 high school mathematics teachers participate in this programme annually.

In addition, BCB participates in other interventions when needs are identified. Examples include the Schools Strategy Workshop intervention; an intervention titled Parents and Schools: Roles and Responsibilities; and arranging school excursions to various companies around South Africa.

BCB conducted an informal survey in 2017 to determine the impact of its efforts since inception to produce key scarce skills qualified graduates. The results are encouraging and impressive:

- 129 students in engineering including the electrical, civil, mechanical, instrumentation and metallurgical fields
- 183 students in commerce, including chartered accountants, economists, auditors and bankers
- 273 students in health and science, including chemists, pharmacists, nurses, doctors, geologists and pathologists
- 17 students in the legal field, including attorneys.

Figures 2.4, 2.5, and 2.6 are published courtesy of the BCB.

Fig. 2.4 Teachers participating in a train-the-trainer programme as part of the BCB

Fig. 2.5 High school students at the BCB

Fig. 2.6 High school students at an opening ceremony of a BCB event

2.3 Institution of Engineering and Technology—Engineering Education Grant Scheme

A strategic priority of the Institution of Engineering and Technology (IET) is demonstrating the public relevance of engineering to society and as a positive programme of study and career choice. IET is emphasising this priority by working in partnership with key stakeholders to promote engineering as a positive career choice, encourage diversity in engineering, provide positive hands-on engineering experience and inspire practising engineers. This strategic priority is clearly in line with IET's stated mission, which is "To inspire, inform and influence the global engineering community, supporting technology innovation to meet the needs of society" (The Institution of Engineering and Technology 2018).

The Engineering Education Grant Scheme (EEGS) was borne out of the merger of the educational grant schemes of the IET and the Institution of Mechanical Engineers (IMechE). This merger was motivated by both organisations' need for better funding consistency for formal and informal learning.

Principally the EEGS focuses on projects within the UK that foster greater understanding of engineering and engineers and their related impacts and contributions. The key target audience are youngsters between the ages of five and 19. There are two levels of grant funding, including programmes and events designed to increase students' knowledge of engineering as a career choice and as a programme of study. The additional funding seeks to enhance "wider engineering literacy", and thus focuses on projects that can further this goal. This may include projects that would have an impact nationally on engineering practice and may also have a continental impact.

Organisations able to participate in the scheme include those that provide STEM activities within the UK. This can include "schools, science communicators, youth clubs, science festivals, museums, science centres, STEM based companies, FE colleges, Higher Education Institutions and members of the IMechE and IET" (IET 2018).

It is important to note that applications to the EEGS must be either led by members, have members as co-applicants or have a member advocate. Demonstrating the involvement and support of local engineers is an important criterion.

The EEGS strongly encourages partnerships wherever possible and is particularly interested in project applications with a multidisciplinary focus involving diverse skill sets and sectors.

The involvement of teachers and students in EEGS is impressive. Since 2015 more than 7100 pre-university teachers and nearly 184,000 students have participated in more than 100 projects.

2.4 Society of Automotive Engineering International—A World in Motion

The Society of Automotive Engineering (SAE) International's A World in Motion (AWIM) has been helping pre-university teachers and their students for more than 25 years. AWIM seeks to enliven the STEM education classroom experience for students aged 5 to 14. It utilises age-appropriate hands-on activities to convey insights that help reinforce concepts learned in the classroom. AWIM aligns with SAE International's mission, which is "To advance mobility knowledge and solutions for the benefit of humanity" (Society of Automotive Engineers International 2018).

AWIM helps students be better prepared for tomorrow's community by helping them to comprehend and deploy STEM notions in complex and practical settings. AWIM also teaches them to increase technological literacy in STEM subjects, learn how to solve complex problems, communicate clearly, ask questions, assimilate information and engage in group activities in order to achieve set objectives. AWIM's principle is to involve and enthuse learners to pursue STEM disciplines as a programme of study and career choice.

Each AWIM activity consolidates educational modules built around the "Designing Plan Encounter" and entails learners working in groups to illuminate a "challenge" to plan, construct and test an item. In expansion, "an industry volunteer" works within the classroom to help instructors with AWIM substance conveyance and serve as a mentor, civic contact and proficient asset. This amalgam of academically sound, standards-based educational programmes, educator professional development, and the use of STEM experts in the classroom has been demonstrated to be viable.

Challenges for students aged five to eight include (Society of Automotive Engineers International 2018):

- Rolling Things: Students learn how to adjust ramp height and vehicle weight to control the momentum of toy cars in the Rolling Things Challenge. Ramp height influences the car's speed when it approaches the crash box; the higher the ramp height, the lower the car's speed when it reaches the ramp's edge. In this unit, the principles covered include inertia, potential and kinetic energy, friction, momentum, weight, velocity, and acceleration.
- Engineering Propelled by Nature: Students examine seeds scattered by wind in the engineering challenge Inspired by Nature. To make paper helicopters and parachutes, they apply what they have learned. To see how these factors affect results, they test different variables (lengths, width, weight, etc.). In this module, the principles covered include inertia, drag, friction, velocity, and acceleration.
- Making Music: In the most up-to-date Making Music Challenge, students investigate sound and vibrations. They learn how the human eardrum works and investigate concepts such as pitch and longitudinal and transverse waves. They collect data through hands-on lessons and design a melodic instrument satisfying specific criteria. A student reader brings the concepts to life for other students through an anecdotal story about creatures and sounds inside nature.
- Straw Rockets: In the Straw Rockets challenge, students investigate the early life of Dr Robert Goddard while perusing the history, "The Rocket Age Takes Off". After examining Goddard's early trials, students construct, test and adjust rockets made from drinking straws. They test the rockets to determine the distance they can fly. Concepts secured in this unit incorporate Newton's Laws of movement, gravity, thrust, lift and drag.
- Pinball Creator: In the Pinball Creator Challenge, students construct, test and adjust a non-electronic pinball machine to make a toy that meets certain specifications. Concepts secured in this unit incorporate gravity, potential and motor vitality and slanted planes.

Challenges for students aged 9–12 include (Society of Automotive Engineers International 2018):

- Gravity Cruiser: Student teams design and build a vehicle powered by gravity. A weighted lever connected to an axle by a string rotates on its fulcrum. This causes the axle attached to the string to turn and propels the cruiser forward as the weight descends. Potential and kinetic power, friction, inertia, momentum, size, length, calculation, graphing and model design are the principles discussed.

- Jet Toy: Students make balloon-powered toy cars that meet particular execution criteria such as voyage distance, weight carried, or speed. Jet propulsion, friction, air resistance and design are the key technological theories learners explore in this challenge.
- Skimmer: Students develop paper sailboats, test the impact of diverse sail shapes and sizes, and develop strategies to meet specific execution criteria. Contact, powers, the impact of surface zone and plan are a few of the physical wonders students experience in this challenge.

Challenges for students aged 12–14 include (Society of Automotive Engineers International 2018):

- Motorised toy car: Students create new designs for electric gear-driven toys. Students are involved in writing plans, making drawings, and working with models to develop a plan to meet specific criteria for layout. The key science concepts addressed by this challenge are force and friction, basic devices, levers and gears, torque and structure.
- Glider: In this challenge, learners investigate the connection between "force" and "motion" and the impact of "weight" and "lift" on a glider. Furthermore, learners study the correlation between manipulating data variables and its analysis. This challenge ends in a public presentation of the designed archetype to the public.
- Fuel Cell: Employing a Proton Exchange Membrane (PEM) fuel cell as the essential control source, student groups plan, construct, and test model vehicles that they must present to a gathering of people. The AWIM Fuel Cells Challenge requires students to investigate physical science concepts such as force, friction and energy transformation, as well as natural concepts such as a green plan, and consolidates arithmetic concepts as they collect, analyse and show data.
- Gravity Cruiser: Student groups plan and construct a vehicle that is fuelled by gravity. A weighted lever associated with a hub by string pivots on its support; as the weight plummets it causes the hub connected to the string to turn, moving the cruiser forward. Concepts investigated incorporate potential and active vitality, contact friction, grinding contact, dormancy, force, breadth, circumference, estimation, graphing, and building a prototype.
- Cybersecurity: In a challenge titled Keeping Our Networks Secure, students explore physical models that simulate the movement of information through the internet. They identify problems with each model and test different enhancements to help make the network operate more efficiently. After learning about the important attributes of cyber security, students work in teams to create marketing materials to help educate various audiences about internet security.

SAE conducted a longitudinal study of the effectiveness of AWIM, focusing on teachers, volunteers and students. This 2005 study concluded that teachers as well as industry volunteers perceived the AWIM programme as positively affecting student learning and interest in STEM-related topics. Teachers reported feeling more confident and more prepared to teach science after using AWIM and volunteers reported that they benefitted from working with teachers and students and that their involvement added value to the learning experience for both teachers and students.

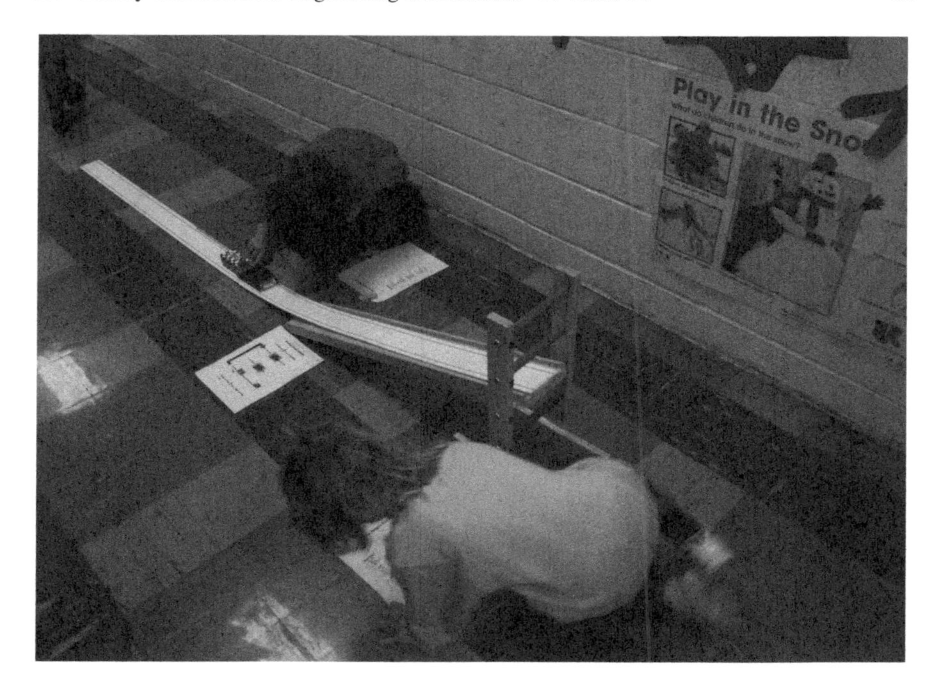

Fig. 2.7 Participants at A World in Motion event

SAE International's AWIM programme was established more than 25 years ago. In that time the programme has grown at an impressive rate. AWIM is now presented in 15 countries and has involved more than 25,000 industry volunteers. These volunteers have helped to influence more than 100,000 pre-university teachers and more than 5 million pre-university students.

Figures 2.7, 2.8 and 2.9 are published courtesy of the SAE.

2.5 Institute of Electrical and Electronics Engineers—Teacher in-Service Programme

Gorham (2006) describes the disconnect between the pre-university education community and engineering and engineers: "Although pre-university teachers, students and parents make daily use of tools and devices that were created by engineers, like cellular phones and iPods, they tend to miss the connection between technology and the people who made this technology happen."

The IEEE has had a long history of promoting and supporting engineering and engineering education at the pre-university level. The IEEE's vision statement exemplifies this focus (Institute of Electrical and Electronics Engineers, Inc. 2018):

Fig. 2.8 Participants at A World in Motion event

Fig. 2.9 Design testing at A World in Motion event

> We will be essential to the global technical community and to technical professionals everywhere and be universally recognized for the contributions of technology and of technical professionals in improving global conditions.

In addition, two of the IEEE's six core values reflect the importance of focusing on pre-university education and the public (Institute of Electrical and Electronics Engineers, Inc. 2018):

> **Growth and Nurturing**: encouraging education as a fundamental activity of engineers, scientists, and technologists at all levels and at all times; ensuring a pipeline of students to preserve the profession.
>
> **Service to Humanity**: leveraging science, technology, and engineering to benefit human welfare; promoting public awareness and understanding of the engineering profession.

The IEEE Educational Activities Board (EAB) is responsible for developing, implementing and coordinating pre-university education programmes to enhance cognizance of the engineering profession and increase of scientific literacy levels amongst pre-university instructors and learners. These programmes are designed to help assure that high-quality science, mathematics, engineering and technology educational backgrounds are available for all pre-university teachers and their students.

To achieve these goals, IEEE has fostered collaboration between engineers and educators through the IEEE Teacher In-Service Programme (TISP), which highlights IEEE members' creation and demonstration of technologically oriented subject matter to local pre-university teachers in an in-service or proficient advancement setting. TISP permits IEEE volunteers to propagate their specialised mastery and to illustrate the application of engineering theories to support the instructing and teaching of science, mathematics and innovation subjects. The focus on nearby schools and school locales is a key highlight of this programme, as engineers and teachers can create long-enduring collaborations.

The objectives are (Institute of Electrical and Electronics Engineers, Inc. 2018):

- Improving pre-university educators "level of technological literacy";
- Establishing engineer/educator relationships to encourage integrated inquiry-based learning; and
- Exposing teachers to career options for their students in engineering and other technical fields.

Since 2001 IEEE's TISP has been helping pre-university teachers, and ultimately their students, to include engineering and engineering design concepts in their teaching creatively each year. To date IEEE members have helped to train over 10,000 pre-university teachers who interact with more than one million students annually.

Of special note are two success stories showcasing how IEEE members in Africa have had a nationwide impact on pre-university education in their country.

(1) After the 2006 TISP training workshop, IEEE members in South Africa were invited to train more than 600 teachers across the country each year. The new curriculum (2006) for Engineering and Technology revealed the need for teachers

to receive training in engineering and engineering principles. These challenges presented an opportunity for IEEE TISP to educate teachers on technical subject matter. Subject specialists from the Department of Education and the nine provincial education departments reviewed and adapted the TISP lesson plans to align with South African education standards, which were demonstrated at a TISP training workshop in August 2006. Other national engineering associations, such as the Engineering Council of South Africa, the South African IMechE, the South African Institution of Civil Engineers, the Institute of Professional Engineering Technologists, and the Chamber of Engineering Technology, cooperated in offering the training sessions.

Further discussions were held, which led to additional training sessions. In 2007, 150 curriculum implementers, subject advisors, and lead teachers were trained. Since then, IEEE members across South Africa have continued to help the South African Department of Education with the development of lesson plans for the South African Technology General Education and Training curriculum.

Equally significant, IEEE members in South Africa were invited by the National Ministry of Education to participate in the rewriting of the engineering and technology curricula for students aged eight to 18. As a result, IEEE members are influencing the teaching and learning of students aged eight to 18 in South Africa currently and for generations to come.

(2) On 28 June 2018 the IEEE and the Tunisian Ministry of Education signed a memorandum of understanding to support IEEE's TISP. The objective is to enhance the quality of teaching by offering new ways of teaching and learning, using engineering and engineering design concepts.

TISP acts in the role of a professional development provider for educators in their bid to introduce practical engineering content into the classrooms. After training, these pre-university educators often introduce this content into their teaching sessions within their local communities (Figs. 2.10, 2.11 and 2.12).

2.6 Summary

Pre-university engineering education and public awareness of engineers and engineering are receiving increased attention worldwide. UNESCO's 2010 report, "Engineering Issues, Challenges, and Opportunities for Development" (United Nations Educational, Scientific and Cultural Organisation 2010), challenged the engineering community and policymakers to develop programmes and activities that address these needs.

This chapter has highlighted the efforts of several engineering societies who have developed and are implementing programmes, competitions and activities designed to increase the public's awareness of engineers and engineering, as well as influencing pre-university students and their teachers.

Through the efforts of these societies more than one million pre-university students and tens of thousands of teachers are enabled each academic year to support and promote engineering as a programme of study and career choice.

Fig. 2.10 IEEE members participating in an IEEE TISP training programme

Fig. 2.11 IEEE members participating in an IEEE TISP training programme

Fig. 2.12 IEEE members participating in an IEEE TISP training programme

References

de Vries MJ, Gumaelius L, Skogh I-B (2016) Pre-university engineering education. Sense Publishers: Springer, Berlin

Gorham D (2006) Work in progress: a pre-university engineering education partnership with engineering societies, educator associations and industry. Paper presented at the 36th Annual Frontiers in Education conference: "borders: international, social and cultural, 28–31 Oct 2006, San Diego, CA. ASEE/IEEE

Institute of Electrical and Electronics Engineers, Inc. (2018) https://www.ieee.org. Accessed on 8 June 2018

Nel H, Ettershank M, Venter J (2016) AfrikaBot: design of a robotics challenge to promote STEM in Africa. In: 19th international conference on interactive collaborative learning (ICL2016) 21–23 Sept 2016, Belfast, UK

Purzer S, Strobel J, Cardella ME (2014) Engineering in pre-college settings: synthesizing research, policy, and practices. Purdue University Press, West Lafayette

Sjaastad J (2010) The influence of parents, teachers and celebrities in young people's choice of science in higher education. Paper presented at the XIV. IOSTE Symposium, Slovenia, 13–18 June 2010, Bled, Slovenia

Society of Automotive Engineers International (2018) https://www.sae.org/. Accessed on 25 July 2018

South African Institute of Electrical Engineers (2018) https://www.saiee.org.za/. Accessed on 20 May 2018

The Institution of Engineering and Technology (2018) https://education.theiet.org/funding/funding-schemes-and-bursaries/. Accessed on 25 May 2018

United Nations Educational, Scientific and Cultural Organisation (2010) UNESCO report: engineering: issues challenges and opportunities for development. Paris, France

Chapter 3
How Professional Societies Support University Engineering Education: Direct Classroom Impact

Abstract Professional engineering societies support engineering education through direct engagement programmes and activities. This chapter highlights how engineering societies are involved in two areas of direct engagement that have an impact on the engineering education classroom: the development and education of industry and programme-specific standards and ensuring programme quality via accreditation. Professional engineering societies are engaged in social innovation programmes and activities that support and promote engineering and engineering education. These programmes/activities have a significant impact on educators, students, practitioners, and the public. There are various categorisations of these activities and their corresponding impact on engineering education. A key categorisation prevalent in literature, is classification based on "classroom" impact (NAE 2017). Thus, engineering societies' engagements are categorised as either formal or informal engagements (NAE 2017). Examples of formal engagements include the development of field-specific standards, involvement in programme accreditation, establishment of student chapters and outreach, as well as national and international competitions. It has also been argued that financial support of students, which most societies do, can be classified as formal support, as it aids students in achieving their educational objectives. Examples of informal (i.e., non-classroom) educational activities include service learning and community service projects, outreach to high schools etc. The philosophy behind the classification hitherto described will be used in this book. However, the terminology we use in our classification is having either a direct classroom impact or an indirect classroom impact. This chapter focuses on examples of engineering societies that have a direct classroom impact, while the following chapter (Chap. 4) discusses activities that have an indirect classroom impact. In this chapter, three direct classroom impact mechanisms will be discussed among the multitude of mechanisms available.

D. Gorham and N. Nwulu, *Engineering Education through Social Innovation*, Lecture Notes in Networks and Systems 108, https://doi.org/10.1007/978-3-030-39006-8_3

3.1 Incorporating Industry Standards into Engineering Curricula

Technical standards are documents drafted with the sole aim of formalising procedures, practices, actions or measures to be strictly adhered to by engineering industry practitioners. Standards can also detail the quality and safety specifications of a product, system or service. The aim is to ensure uniformity of engineering procedures, systems or services and hence ensure safety and reliability. Depending on the context or jurisdiction, standards may either be voluntary or compulsory. Accreditation bodies such as the American National Standards Institute (ANSI) exist to define modalities for standard-designing organisations to follow when developing standards. There has also been a recent drive to include standards education and teaching about standards and standardisation in undergraduate curricula. This was initially spearheaded by the Accreditation Board of Engineering and Technology and has spread to most universities across the world. This drive initially focused on familiarising students with the technical specifications of various standards but has shifted towards also familiarising students with the development process of standards. Thus, students who understand this process would be an asset to employers and would be readily involved in future standards designs. A comprehensive listing of successful implementations of technical standards in engineering undergraduate education is given by Lemeš (2015) and Olshefsky (2008). Furthermore, a framework for incorporating standards into engineering curricula is detailed by Sinha (2010).

Engineering societies that are actively involved in introducing standards into undergraduate engineering education curricula are detailed below:

3.1.1 *American Society of Mechanical Engineers*

The American Society of Mechanical Engineers (ASME) has been involved in the development and implementation of industry standards since the organisation was founded in 1880. It currently has over 600 standards and codes, covering a variety of technical areas. ASME has published a booklet for university students entitled, "ASME: Standards and Certification: Examples of Use of Codes and Standards for Students in Mechanical Engineering and Other Fields" (American Society of Mechanical Engineers 2008). This booklet includes articles focusing on the many aspects of ASME's standards and codes and includes an overview of how standards and codes are developed. The section on the development of standards and codes focuses on:

- the need for industry standards and codes;
- explaining what a standard is;
- providing an outline of the process of how ASME develops standards and codes.

As part of their education, university students are given the opportunity to gain knowledge and insight regarding the crucial area of industry standards through the efforts of ASME.

3.1.2 Institute of Electrical and Electronic Engineers

Globally, the IEEE represents a top standards developer in technological methods for disseminating information and resources on principles, applications required in designing new products, methods, and services. This becomes important for students in tertiary institutions, both the ones currently studying and the graduates in engineering fields that have little or no experience of the standards required in their course of study. Students in the engineering field become valuable employees to their various employers when they possess essential understanding of technical standards.

With the identification of the significant role that standard plays in today's world of work, the IEEE has committed resources to educate and enlighten students, professionals, academics in engineering, technology, and computing on the importance of standards in their various fields. This is believed to enhance the transition of the students' knowledge of standards to become professionals who effectively put this knowledge into practice in the real world.

Furthermore, the recognition of technical standards is of great benefit not only to engineering students but also to students in fields such as economics, business, political science and law. Research has shown that during their undergraduate studies, students are not adequately exposed to the technical standards that have influenced several disciplines in educational institutions. Accordingly, to facilitate understanding of technical standards and promote their incorporation into academic programmes, the IEEE has developed education programmes in standards, provided learning materials and resources for university students, practising engineers and academics.

The IEEE has developed the IEEE Standards University as a delivery mechanism to fulfil its strategic commitment to:

- Support the significance of technical standards in addressing the challenges of economic, technical and environmental factors;
- Provide learning materials required for the application of technical standards in the design and development of educational programmes;
- Design short courses on technical standards related to the design and development stages in professional practice; and
- Promote the incorporation of technical standards into academic programmes.

IEEE Standards University greatly expands the contents and resources of IEEE standards for students, academics and professionals. Below are brief descriptions of some of the features of StandardsUniversity.org (Institute of Electrical and Electronics Engineers, Inc. 2018):

- IEEE Standards Education e-Magazine;
  - This online publication is designed for university students, faculty and practising engineers.
  - The aim of this publication is to highlight and demonstrate the significance of standards applications required in the process of standard developments to readers. The knowledge of this application will assist undergraduate students in higher institutions and young professionals to facilitate their career in engineering and technological development to benefit future generations.
- IEEE Standards Education Videos.
  - Videos have been developed for university students, young professionals, faculty, practising engineers and industry executives. Sample titles include:
    The transition from university: standing out with standards.
    Advance your career: standing out with standards.
    Educating students on standardisation in engineering, science and technological studies.
    Standards insights for executives: A five-part series.
    The design system of rescue robots: Standard lays the foundation from prototypes to deployment.
    Develop a global business sense: Standing out with standards.
- IEEE Standards Simulation Game
  - Creating a Mars Space Colony is the first requirement for educators, students and professionals to develop effective learning methods, working in teams and negotiating and reaching building consensus. This is a game of standardisation that involves a standards development simulation game created by experts with wide experience in high-stakes, real-world technical standard development.
  - This virtual game is intended to teach technical and development standards and at the same time to develop skill in negotiation and teamwork. As a result, topics such as the fundamentals of standard development and importance of standards, as well as a case study on standards, should be included in the teaching curriculum.
    As members of standard working groups, players contribute by incorporating the roles that reflect the economic, political and technical realities of standards development.
  - Participation in the simulated game does not require a technical background.
- IEEE Standards Education Student Grants Programme.

IEEE Standards Education Grants are offered for students to help students and graduates design projects and research in which industry technical standards are applied. An honorarium for faculty mentors is part of the grant programme. At the end of the project, students produce a student application paper. If accepted, the paper is published on the IEEE Standards Education website.

Table 3.1 Engineering societies and their corresponding BOK

	Engineering Society	Body of Knowledge Developed
1	IEEE Computer Society	Software Engineering Body of Knowledge
2	The International Council on Systems Engineering, Institute of Electrical and Electronics Engineers, and Systems Engineering Research Centre	Guide to the Systems Engineering Body of Knowledge
3	The ASME	Guide to the Engineering Management Body of Knowledge
4	The National Society of Professional Engineers	Professional Engineering Body of Knowledge
5	The American Academy of Environmental Engineers and Scientists	Environmental Engineering Body of Knowledge
6	Society for Manufacturing Engineers	Body of Knowledge for Certified Manufacturing Engineering Programmes

3.2 Design and Development of Engineering Body of Knowledge

The American Society of Civil Engineers (ASCE) has spearheaded a significant contribution in engineering education by conceptualising and detailing the Body of Knowledge (BOK) project (American Society of Civil Engineers 2008). The BOK is a clear and succinct description of the prerequisite knowledge and skills that civil engineers should have. This knowledge and these skills are further subdivided into four foundational, 11 technical, and nine professional outcomes. The concept of BOK has been well received and has been a significant influence in other sister and sub-disciplines (Varma 2005; Theis et al. 2008).

A list of engineering societies that have created their own BOK is shown in Table 3.1.

3.3 Ensuring Programme Quality Through Accreditation

The International Engineering Alliance (IEA), comprised of leading engineering institutions from 23 countries, is an umbrella organisation for seven multilateral agreements that have created international benchmark standards for education in engineering. Member organisations enforce these standards to assure programme quality in engineering education and competence to practise in an engineering field at the entry level. These seven agreements/accords include the agreement of the Asia-Pacific Economic Cooperation, International Professional Agreement, Agreement for International Engineering Technicians Agreements, International Engineering Technologist Agreement and the Sydney, Washington, and Dublin Accords.

IEA's 2014 publication "25 Years of the Washington Accord: 1989–2014 Celebrating international engineering education standards and recognition," (International Engineering Alliance 2014) notes that the 15 signatories of the Washington Accord, namely the UK, South Africa, the US, Australia, Singapore, Korea, Canada, Ireland, New Zealand, Hong Kong China, Japan, Chinese Taipei, Malaysia, Turkey and Russia, have recognised the need to ensure that engineering graduate attributes include a strong technical foundation, the technical and professional skills these graduates should demonstrate and the attitudes they should possess to contribute to and lead in today's global environment.

Hoosain and Sinha (2018) note the challenge that colleges of engineering face in keeping curricula current:

> Engineering educators at universities throughout the world are challenged to review curricula continually to prepare students to succeed in the ever-changing professional world. Universities are under pressure to produce graduates who can deal with short-, medium- and long-term societal challenges in a more relevant way. The relevance component has often required universities to graduate students whose skills are locally relevant, yet internationally competitive. Technologies will always change, therefore developing students' ability for life-long learning is much more necessary today than previously. (p. 2)

The aim of programme accreditation is to certify degree-granting academic programmes. The accreditation is a formal recognition of a programme from a third party, which specially evaluates the quality of this programme with respect to an agreed-upon set of reference educational standards. The motive is for relevant stakeholders to evaluate the programme's quality and the degree to which it abides by prerequisite standards. The relevant stakeholders can include employers, students, alumni, industries amongst others. The accreditation process assists educational institutions in self-introspection, which leads to continuous programme improvement and effective stakeholder engagement on curriculum and pedagogy. Accreditation also serves an added advantage of indicating if graduates of a program received training at a level that makes them immediately employable or grants them candidate status, which enables them to register later as professionals. Programme accreditation helps in providing quality education, better career and international mobility options to the students, well-qualified and skilled manpower to the industries, and reputation and visibility to educational organisations.

In many countries programme accreditation is carried out by a third party, typically non-governmental. Examples include the Accreditation Board for Engineering and Technology (US) and the Engineering Council of South Africa. Most professional engineering societies support programme accreditation and participate in it through the accrediting body by supplying evaluators, offering technical expertise and policy guidance. This is a key influence of professional societies on engineering education.

The next section describes how IEEE has approached programme accreditation on a global level, leveraging local members.

3.3.1 IEEE's Approach to Programme Accreditation

The IEEE favours the creation of local, regional, and multinational bodies and there is full representation of constituents in countries where the programmes are certified. IEEE promotes and supports the advancement of common recognition of agreement between accrediting agencies. The policy of IEEE is to support regional or country accreditation when requested and where possible. The aim of the regional and country-based accredited programmes is to give opportunities to stakeholders who teach, learn and carry out business, and implement the accreditation criteria. The IEEE's involvement in global accreditation also serves to further the future of the profession and helps the IEEE stay current with university education issues on a global scale. When requested, the IEEE through its appropriate member board and committees provides guidance, support and resources to facilitate the creation of regional/country accrediting bodies and related services.

One such example of the IEEE playing a key role in forming a new accrediting body is the creation of the Instituto de Calidad y Acreditación de Programmeas de Computación, Ingeniería y Tecnología (ICACIT) in Peru (ICACIT 2018).

ICACIT is a non-profit organisation, founded in 2003, comprising five professional and technical societies working collaboratively to improve the criterion of accreditation to ensure quality higher education and continuous improvement centred on computing, engineering and engineering technology programmes in Peru.

The objective of ICACIT is to contribute to national progress through quality higher education. To this end, ICACIT supports and promotes programme accreditation and the continuous improvement of engineering, computing, and engineering technology programmes in Peru.

In addition, ICACIT has a long-term goal of worldwide recognition of its programmes in engineering, computing and engineering technology. As a non-profit NGO, independent of the institutions offering engineering and computing programmes and substantially led by industry, ICACIT has the support of various stakeholders in the engineering profession to be the sole applicant to membership to the Washington Accord representing Peru.

During its formation and establishment as an accrediting body the IEEE, through its members, assisted ICACIT by:

- organising and sponsoring informational workshops for stakeholders from academia, industry and government designed to describe programme accreditation, its benefits, the costs, the process and the work involved;
- providing training workshops for programme evaluators;
- advising the organisers regarding organisational structure, financial considerations, governance and administration;
- leveraging members and professional staff to assist in developing appropriate documents, e.g., governance, accreditation model, self-study;
- providing financial, member and professional staff support for the informal visits of identified programmes;

- providing member and professional staff support in developing and maintaining a web presence;
- providing financial, volunteer and professional staff support for observers to participate in accreditation visits in other countries, e.g. the USA; and
- providing guidance on developing the "story" of accreditation and delivering this story to academia, industry and government

Since the start of the journey to establish ICACIT as an accrediting body in Peru, several significant milestones have been reached, including:

- 2007–2008: ICACIT conducted its first accreditation visits.
- 2014: Admission to the Washington Accord (WA) as provisional signatory.
- 2016: Admission to the Sidney Accord (SA) as a provisional signatory.
- 2016: SINEACE (National Accreditation System in Peru) recognises accreditations granted by ICACIT.

Today ICACIT continues to expand its work in programme accreditation in Peru. It has accredited more than 40 programmes and 15 reaccredited programmes at 14 universities to date. It has more than 65 active programme evaluators and recruits 10–15 new programme evaluators each year. It conducts programme evaluator training multiple times per year and has a cadre of eight training facilitators.

ICACIT was recognised as a full signatory to the Washington Accord in June 2018 and is expected to become a full signatory to the Sidney Accord in 2019–2020.

3.4 Chapter Summary

This chapter has described several social innovation strategies that have a direct impact on the engineering education classroom. From (a) assuring engineering education quality through programme accreditation, including the creation of new accrediting bodies, to (b) developing academic BOK providing a clear and concise description of the prerequisite knowledge and skills for various engineering programme areas and (c) developing a variety of strategies and activities to incorporate technical standards into the engineering curriculum, engineering societies are making significant contributions to engineering and engineering education.

References

American Society of Civil Engineers (2008) Civil engineering body of knowledge for the 21st century: preparing the civil engineer for the future. Body of Knowledge Committee, Reston, VA

American Society of Mechanical Engineers (2008) Examples of use of codes and standards for students in mechanical engineering and other fields. https://www.asme.org. Accessed on 15 July 2018

Hoosain MS, Sinha S (2018) Integrating 'Engineering Projects in Community Service' into engineering curricula to develop graduate attributes. SOTL South 2(1):60–75

Institute of Electrical and Electronics Engineers, Inc. (2018) https://www.ieee.org. Accessed on 15 June 2018

Instituto de Calidad y Acreditación de Programas de Computación, Ingeniería y Tecnología ICAICT (2018) https://icacit.org.pe/web/eng/. Accessed on 15 May 2018

International Engineering Alliance (2014) 25 years of the Washington Accord. Wellington, New Zealand

Lemeš S (2015) Standardization in engineering curriculum. In: 9th Research/Expert conference with international participations, Neum, B&H, 10–13 June 2015

National Academy of Engineering (2017) Review of the literature on engineering society involvement in undergraduate engineering education. National Academy of Sciences, The National Academies Press, Washington, DC, USA

Olshefsky JP (2008) The role of standards education in engineering curricula. In: American society of engineering education's fall 2008 ASEE conference, Mid-Atlantic Section

Sinha S (2010) The role of technical standards in engineering, technology and computing curricula. IEEE Position Paper, 2010

Theis R et al (2008) Pathways to learning: orchestrating the role of sustainability in engineering education. In: 2008 ASEE annual conference and exposition. American Society for Engineering Education

Varma VK (2005) Basic elements of the 21st century body of knowledge for a construction professional: challenges for construction educators. In: 2005 ASEE annual conference and exposition. American Society for Engineering Education

Chapter 4
How Professional Societies Support University Engineering Education: Indirect Classroom Impact

Abstract Engineering societies support university students via indirect engagement programmes and activities. This chapter focuses on examples of indirect engagements developed and entered into by professional engineering societies that are promoting and enhancing the content and delivery of engineering by focusing on and affecting university students and the public. Three indirect engagement mechanisms will be discussed: Engineering Projects in Community Service in IEEE; the Society of Automotive Engineers Collegiate Design Series; and the American Society of Mechanical Engineers (ASME) E-Fests. The concept behind these activities is to foster a passion in students for engineering design activities away from the curriculum and conventional classroom. These activities serve as an avenue for students to use the skills obtained in the classroom to deal with practical challenges, while also enhancing their teambuilding ethos.

4.1 Engineering Projects in Community Service in IEEE

Engineering Projects in Community Service (EPICS) in IEEE is a programme that oversees high school students, university students, engineering faculty and staff from NGOs to work on community-based projects in engineering (Institute of Electrical and Electronic Engineers, Inc. 2018). This offers students the exceptional opportunity to be involved in their local communities, discover their interests and career choices, and benefit from the real professional experience with hands-on engineering and technology design projects for their community. Globally, EPICS in IEEE is changing lives with an exceptional mix of technology and education. Ever since its 2009 launch, the programme has become a top global resource for engineering students and engineers in search of technological solutions and support to solve local community problems. EPICS in IEEE has proven to be an effective tool for recruiting girls and young women to technological careers; one-third of all university students working as volunteers in this programme are female, an underrepresented group in the STEM professions. The concept of collaboration of students with agencies of community service to resolve problems in engineering started with EPICS at Purdue University's College of Engineering.

D. Gorham and N. Nwulu, *Engineering Education through Social Innovation*, Lecture Notes in Networks and Systems 108, https://doi.org/10.1007/978-3-030-39006-8_4

Today, EPICS in IEEE is an established platform to make an instant impact and offer long-term sustainability by building intellectual capital, technological literacy and human resources planning (Institute of Electrical and Electronic Engineers, Inc. 2018). Young adults at a key stage of their development stand to benefit from an inestimable, life-shaping experience when they volunteer to work with EPICS in IEEE. Furthermore, to stimulate a lifelong interest in community engagement, university undergraduates or high school students participate in the programme to develop first-hand appreciation for how technology advances their lives in their communities. The knowledge acquired by university students as EPICS in IEEE volunteers will go a long way to nurture and help to develop their future careers. Significant opportunities offered to widen the skills of undergraduate students are made available through the process of defining, designing, building, testing, deploying and supporting an engineering-based solution.

This is important for upcoming engineers who desire to explore the professional world, where hard and soft skills are required to excel, in addition to technical know-how in their profession.

Skills such as leadership, communication project management and teamwork are not criteria in the engineering curricula. However, these skills are vital for a successful career in an ideal professional work environment. Several opportunities are offered to student volunteers taking part in EPICS in IEEE projects to develop these skills.

In much the same way, high school students also benefit from participating in EPICS in IEEE. The programme helps them build knowledge and self-confidence, while at the same time receiving a stimulating, hands-on introduction to STEM as a programme of study and career choice.

In 2014 IEEE contracted with McKinley Advisors to conduct a study to evaluate the impact of five EPICS in IEEE projects. Seventy-five percent of the university students who participated said the greatest benefit of taking part in EPICS in IEEE was learning how engineering can solve real-world issues. All university students agreed that EPICS in IEEE had helped to enhance their ability to understand the needs of end-users, as well as enabled them to put theoretical knowledge into practice. To enable the implementation of solutions to challenges faced by communities, EPICS in IEEE allows students to use technical knowledge acquired to work with local service organisations. In essence, EPICS in IEEE not only benefits the communities, but also promotes community improvement through career development of students in engineering.

The strength and appeal of the EPICS in IEEE programme have attracted the attention of the IEEE Foundation, which has identified EPICS in IEEE as a programme that highlights the possibilities of technology in addressing global challenges. EPICS in IEEE has a track record of furthering the IEEE's main objectives of promoting technological innovation and excellence for the benefit of society.

The EPICS in IEEE programme links engineering with community service in four sets of community improvement efforts:

Access and Abilities

- There is a better capability to resolve accessibility concerns in communities by linking student branches at universities, secondary students and non-profit organisations.
- Research conducted by engineering students helps to develop their communities.
- Access and abilities projects initiated by EPICS in IEEE foster adaptive services, healthcare facilities for children with disabilities, programmes for adults and assistive technologies.

Education and Outreach

- EPICS in IEEE program enables young learners to engage with the benefits of STEM disciplines thereby positioning them for future careers in STEM fields.
- Several projects allow students to gain practical experience to promote their interest in these areas. Through these EPICS in IEEE ventures, societies and schools that lack strong engineering programmes are gaining new curricula along with new facilities to explore new topic areas.

Environment

- Science and engineering are crucial approaches to environmental problems. With the evolution of technology and the need for conservation, ecosystems shift in societies in every area around the world. Most EPICS ventures are concerned with innovative ways of generating power and heat, recycling and the use of renewable energy sources in IEEE projects. Young students learn about the effects of environmental issues and how technology can be part of the solution through these EPICS in IEEE programmes. Students also gain exposure to potential jobs related to the increasing demand for alternative energy and solutions to environmental problems.

Human Services

- Through their encounters with human administration EPICS in IEEE ventures, students discover associations between designing and the colossal scope of community needs universally. This incorporates vagrancy anticipation, reasonable housing, family and children agencies, neighbourhood revitalisation and local government. Even after an EPICS in IEEE project has been completed, its lasting impact continues to be felt through the local non-profit organisation's involvement.

Examples of projects

- Collaborating for educational opportunities

The lights are on for 600 students at a primary school in rural, isolated Kasiluni, Kenya. Because the school is not on the national power grid, a partnership between IEEE Kenya student members, student groups at the Jomo Kenyatta University of Agriculture and Technology and a sustainable energy provider engineered solar-powered electricity. Now, the school is open for longer hours, electronic devices can be used, and overall educational quality has been greatly improved. More importantly, the student members got practical experience installing solar technology, and secondary school students who helped out learned about renewable energy and electrical engineering careers.

- Acquiring a world view

Members of the IEEE STEM (iSTEM) Club at New Jersey's Bridgewater-Raritan High School, USA became aware of the global digital divide while helping students in Paushi, a rural village in India. Partnering with IEEE volunteers and Kreeya, a non-profit organisation, the iSTEM Club used EPICS in IEEE grants to develop a basic library, help install computers, and plan and design a new cyber classroom. A teacher provides basic computer training so Paushi children can prepare for India's employment market. They are learning both digital literacy and English—and iSTEM Club members have new engineering and project management skills.

- Building ecological understanding

The University of New Hampshire's (UNH) Oyster Restoration Programme worked with the Nature Conservancy to help restore the oyster population in New Hampshire's Great Bay. Oysters are a critical part of the ecosystem, filtering out pollutants to help other organisms survive. A team of UNH students and high school students developed a network of electronic water flow meters for measuring the sedimentation rate, which is key to cultivating new oyster beds. In a win-win for all, the UNH students received real-world experience, the high school students learned basic engineering concepts and a local cause received important help.

- Innovating for food security

In La Paz Centro, Nicaragua, a centre of drought and hunger in Latin America, students from Vassar College, the University of Managua and a local high school STEM group are developing an innovative approach to agriculture. Working with the non-profit group Artists for Soup, the multi-disciplinary team is creating 70 bio-intensive garden beds at the high school. Using simple, inexpensive components, the students are also inventing new strategies for water collection and slow-drip irrigation. Once the system is operating, the students will have a better way to produce food, while also growing their knowledge of engineering.

- Developing environmental awareness

Since 2009, a partnership in densely populated South Philadelphia has provided its residents with vital air-quality information. Natural gas extraction in the nearby Marcellus Shale region had raised concerns about the environmental effects of drilling, and few monitoring solutions addressed air quality. The collaboration between IEEE student members at Drexel University, North Penn High School and the Clean Air Council, comprising 11 local lung associations, is an ongoing effort. Student teams continually enhance the design and effectiveness of the sensor networks they created, and the Council has a stronger voice in air-quality regulatory policies.

- Helping others achieve independence

An inexpensive, handheld device promises new dignity and confidence for the speech and hearing-impaired, thanks to IEEE student members at RNS Institute of Technology in Bangalore, India. Since 2012, student teams have worked with educators at a local school for hearing and differently abled children on a system to translate sign language simultaneously into text and audio. When completed, it will bridge the gap between those who cannot speak or hear to those who do, and it is hoped that this will fully open up employment opportunities for users. Meanwhile, the student members have gained valuable insights about innovation and teamwork and have enjoyed the satisfaction of helping others.

- Improving quality of life

In Kathrada Park, home to some of the poorest in Johannesburg, South Africa, UJ students used the Litre of Light solution to provide safe night-time lighting. With high school students, the UJ students designed a system to charge batteries using solar energy. A solar cell is attached to a plastic bottle functioning as a daytime skylight and inserted into a thin tin roof. Now, instead of devastating fires caused by overturned kerosene lamps, safe, sustainable lighting is making life a little better in Kathrada Park, and students have learned that engineering can also be about caring. The student leader has also developed a scholarly paper (Figs. 4.1, 4.2, 4.3 and 4.4).

4.2 SAE—Collegiate Design Series

The SAE is committed to supporting and encouraging engineering and engineering education and the developing new generation of engineers. The organisation's vision statement says: "SAE International is the leader in connecting and educating mobility professionals to enable safe, clean, and accessible mobility solutions" (Society of Automotive Engineers International 2018).

One of the societies' core goals focuses on education and professional development initiatives for STEM that encourage and create the future workforce of mobility.

SAE International's Collegiate Design Series (CDS) focuses on undergraduate and postgraduate students worldwide participating in a variety of competitions. These

Fig. 4.1 EPICS in IEEE: Designing, building and installing environmentally friendly solar-powered phone chargers and charging controllers to allow low-income communities in Uganda access to the power grid

Fig. 4.2 EPICS in IEEE: IEEE Student members from UJ, South Africa are working with pre-university students from UJ Metropolitan Academy to develop an insect-monitoring system for plant protection to be used by local farmers in the area

Fig. 4.3 EPICS in IEEE: "Waste Electrical and Electronic Equipment Recycling Programme" (National University of Cordoba, Argentina)

Fig. 4.4 EPICS in IEEE: Mobile Microsensors—Air Quality Monitoring

competitions take students past reading material hypotheses by empowering them to plan, construct, and test the performance of a genuine vehicle and after that compete with other students from around the globe in energising and intense competition.

CDS comprises a variety of competitions, including:

- SAE Aero Design—The SAE Aero Design competition is a practical design challenge that seeks to curtail an orthodox aircraft development programme into a single calendar year, allowing participants to break down requirements through the system engineering process. It introduces participants to conceptual design, manufacturing, system integration/testing and sell-off complexities through presentation.

- AutoDrive Challenge—This recently set up, three-year independent vehicle competition, sponsored by SAE International and General Motors, challenges students to create and illustrate a fully independent driving traveller vehicle. The specialised objective of the competition is to explore an urban driving course in a mechanised driving mode, as portrayed by SAE Standard (J3016) Level 4 definition, by the third year.
- Baja SAE—In Baja SAE, design students are entrusted with planning and building a single-seat, all-terrain vehicle that can serve as a model for a dependable, viable, ergonomic, and economical vehicle that could be marketed to recreational clients. The students must work as a group to plan, build, construct, test, advance, and compete with a vehicle within the rules.
- Clean Snowmobile Challenge—The SAE Clean Snowmobile Challenge gives students the opportunity to upgrade their design plan and extend administration aptitudes by re-engineering an existing snowmobile to diminish emissions and noise. Participants' adjusted snowmobiles compete in an assortment of events, including reducing emissions and noise, fuel economy/endurance and increasing speed.
- Formula Hybrid—The Formula Hybrid Competition is an intriguing planning and building challenge for undergraduate and graduate college students. They must collaboratively plan and construct a formula-style electric or plug-in hybrid race car and compete in a series of events. This instructive competition emphasises drivetrain advancement and fuel effectiveness in a high-performance application.
- Formula SAE—Formula SAE series contests challenge university and student teams to model, design, produce, create and compete with lightweight, formula-style vehicles. Involving teams from other universities around the world, the competitions give teams the opportunity to demonstrate and prove both their ingenuity and engineering skills.
- Formula SAE Electric—The Formula SAE Electric project allows fully electric vehicles to be built within the framework of Formula SAE. Teams use electric motor-powered vehicles and compete in static and dynamic events, comparing elements such as design, presentation, price, acceleration, skid pads, autocross, stamina and performance.
- SAE Supermileage—SAE Supermileage's engineering design goal was to develop and build a one-person, fuel-efficient vehicle that satisfies the rules of the competition. The designed vehicles need to obtain the highest combined km/L (mpg) rating and students are exposed to a design portion of the work with assessments in the form of a written report and a verbal presentation.

More than 10,000 students take part in SAE International's Collegiate Plan Arrangement (CDS) competitions each year. CDS competitions introduce undergraduate and graduate design students to an assortment of disciplines in mobility-related business by involving them in hands-on group building activities that require budgeting, communication and asset administration. Students are moreover introduced to specialists from driving companies in the industry, which is important when they start working after graduation.

The SAE administers surveys of faculty advisors and participating students to assist in determining the impact of the CDS. Survey results have indicated the following impact of the CDS on participating students:

- 91% of students demonstrated an improvement in leadership skills.
- 96% of students demonstrated an improvement in teaming skills.
- 93% of students demonstrated an improvement in project management skills.
- 92% of students demonstrated an improvement in communication skills.
- 71% of students demonstrated an improvement in finance and budgeting skills.

SAE's CDS programme participation is embraced by the top 100 engineering degree-producing colleges/universities as reported in the American Society of Engineering Education's (ASEE) Profiles of Engineering and Engineering Technology Colleges. At present each of the top 50 engineering degree-producing universities/colleges participates in one or more of SAE's CDS programmes.

4.3 ASME—E-Fest

The ASME has long been involved in supporting and promoting engineering and engineering education. This is reflected in ASME's mission, which states (American Society of Mechanical Engineers 2018): "ASME's mission is to serve diverse global communities by advancing, disseminating and applying engineering knowledge for improving the quality of life; and communicating the excitement of engineering."

ASME's participation in programmes and activities to support and promote engineering and engineering education is aligned with two of the seven core values of the organisation: "Facilitate the development, dissemination and application of engineering knowledge" and "Promote the benefits of continuing education and of engineering education."

Two of the society's goals support a focus on engineering education for current and future generations of engineers: "ASME offers education and training programmes to prepare the workforce of tomorrow to address the world's challenges" and "ASME engages and inspires future generations to pursue careers in engineering."

ASME is providing programmes focused on university students to increase engagement and enhance their education. The society is accomplishing this through engineering festivals (E-Fest). These in-person events provide attendees with the opportunity to increase their engineering knowledge, participate in competitions, demonstrate leadership, participate in student research, meet industry professionals, obtain career advice, understand better how engineering benefits society and celebrate engineering.

E-Fests are held in several countries each year. Locations have included India, Brazil and the US. E-Fests have been held for two years and attract thousands of university students. Each E-Fest typically includes one or more keynote addresses,

Fig. 4.5 Participants at ASME E-Fest West 2018

industry-focused panel discussions, interactive workshops, networking, social activities and student competitions, including the Human Powered Vehicle Challenge, the Student Design Competition, the Old Guard Competitions and ASME Innovative Additive Manufacturing 3D Challenge (Figs. 4.5, 4.6, 4.7 and 4.8).

4.4 Chapter Summary

This chapter focused on multiple programmes and activities that engineering societies have developed and implemented in support of the education of university students and to enhance the engineering classroom experience via indirect engagements. The programmes and activities described in this chapter focused on: (a) developing a passion in engineering students for engineering design activities outside the traditional classroom; (b) instilling a sense of giving back to their local communities by leveraging technology to solve local problems; and (c) exposing students to needed skill sets, including project management, communication, working in teams, budgeting and leadership. These programmes and activities serve as a path for engineering students to implement the skills obtained from the classroom, focusing on a practical challenge.

Fig. 4.6 Participants at ASME E-Fest Asia Pacific 2018

Fig. 4.7 Participants at ASME E-Fest Asia Pacific 2017

Fig. 4.8 Participants at ASME E-Fest East 2018

References

American Society of Mechanical Engineers (2018) https://www.asme.org. Accessed on 6 July 2018

Institute of Electrical and Electronic Engineers, Inc. (2018) https://www.ieee.org. Accessed on 24 June 2018

Society of Automotive Engineers International (2018) https://www.sae.org/. Accessed on 25 June 2018

Chapter 5
How Professional Societies Support Engineering Education: Informal Education

Abstract Educational programmes and activities outside the formal education system provide learning opportunities for students, teachers and the public in science, technology, engineering and mathematics. Science and technology museums are highlighted, as they are well positioned to provide teachers, students and the public with teaching and learning opportunities in STEM that are locally relevant, promote engagement among participants and reflect the local culture. This chapter will concentrate on the efforts of two engineering societies and their education programmes and activities in collaboration with science and technology museums.

5.1 Why Informal Science, Technology, Engineering and Mathematics Education?

Educational programmes and activities outside formal teaching provide learning experiences for students, teachers and the public in STEM subject areas. Museums, nature centres, zoos, parks, aquariums, radio, television, internet, and other science-rich institutions and/or media are effective venues to reach students, teachers, parents and the public. Science centres are often strategically placed to house and facilitate the meeting of innovators, innovations and end-users.

The US National Research Council (2009) reported that "learning experiences across informal environments may positively influence children's science learning in school, their attitudes toward science, and the likelihood that they will consider science-related occupations." (p. 304) In numerous areas, science museums are the only facility in the community where pre-university instructors, students, and the public are exposed to hands-on experiments in science, technology, engineering and mathematics

This chapter will concentrate on efforts involving engineering societies and their education programmes and activities in collaboration with science and technology museums. Science and technology museums are viewed as local centres for community engagement and learning. This community engagement opportunity allows local units of engineering societies to interact with the local science and technology museums and contribute to their educational programmes. In addition, an opportunity

D. Gorham and N. Nwulu, *Engineering Education through Social Innovation*, Lecture Notes in Networks and Systems 108, https://doi.org/10.1007/978-3-030-39006-8_5

is created for engineering societies to engage with teachers, students and the community and inform them of science, design, innovation and arithmetic and related careers in science and technology museum environments.

Expectations for today's science and technology museums have shifted. Koivu and Myllykoski (2015) describe how science centres have evolved to be an integral part of the local community:

> During the last quarter of a century, new science centres have been founded around the globe at the rate of roughly a hundred per year. As societies become increasingly urban, science centres are being built to meet educational and cultural needs and to strengthen local identities. Science centres play a role in inspiring and motivating audiences, in strengthening scientific knowledge and skills and in reinforcing science's importance and impact in future societies. Science centres have attracted public investment because they demonstrate how science and curiosity are a central part and ingredient of humanity - and of our well-being. Science centres are also (and increasingly) here to act as venues of dialogue and participation in science and its applications. Typically, science centres do not take an active role in formal education. Instead, learning at our venues is free-choice or voluntary – no one fails at a science museum. However, for educational and research institutions, science centres offer a natural channel for dissemination and interaction with wide audiences. To communities and local authorities, science centres are not only attractions, but also hubs for various activities and ways to enrich the lives of local citizens. This has justified public support for many science centres, which rely not only on public funding but also on active co-operation between a variety of stakeholders, many of which are also founders of the centres.

Science and technology museums are well-positioned to provide teachers, students and the public with opportunities and mentoring in STEM that are locally relevant, promote engagement among participants and reflect the local culture. In the US National Research Council's 2015 report, "Identifying and Supporting Productive STEM Programmes in Out-of-School Settings" (p. 5), the concept of a STEM learning ecosystem is described and includes a community's STEM-rich assets, among others:

- outlined settings, such as schools, clubs, museums, and youth programmes;
- non-artificial settings, such as city parks, conduits, and timberlands and deserts;
- individuals and networks of people, such as practising STEM experts, teachers, devotees, enthusiasts, hobbyists, and business leaders who can serve as an inspiration and role models; and
- regular experiences with STEM, such as on the Web, on television, in the play area, or amid discussions with family individuals and other youthful people.

Providing access to engaging and thought-provoking STEM learning opportunities in informal education settings is important in enriching STEM learning for teachers, students, and the public. The US National Research Council (2015, p. 15) report goes on to state that: "productive programmes are intellectually, socially, and emotionally engaging." It is important to add that effective learning opportunities reflect and create the learner's interest in and understanding of STEM and are associated with the broader scope of STEM learning and scholastic and career pathways.

The National Research Council's 2015 report (p. 15) included a synthesis of the research and established criteria for distinguishing effective out-of-school STEM programmes. The criteria fall into three categories:

1. Productive programmes engage young people intellectually, socially, and emotionally.
    - First-hand experiences with phenomena and materials are provided.
    - Youths are engaged in sustained STEM practices.
    - A supportive learning community is established.
2. Effective programmes respond to young people's interests, experiences, and cultural practices.
    - STEM is positioned as socially meaningful and culturally relevant.
    - Collaboration, leadership, and ownership of STEM learning are supported.
    - Staff serve as co-investigators and learners alongside young people.
3. Interface STEM learning in out-of-school, school, domestic, and other settings offers a profitable programme.
    - It connects learning encounters over settings.
    - It leverages community assets and organisations.
    - It actively brokers additional STEM learning opportunities.

Members of engineering societies are often very active in their local communities. Working relationships between members of engineering societies and/or between the local unit of an engineering society and local science and technology museums may exist or can be established.

Below we describe the efforts of two engineering societies and their programmes for teachers, students, and the public, leveraging science and technology museums.

5.2 China Association of Science and Technology: "Bridging Science Museum with School"

The China Association for Science and Technology (CAST) is the largest non-governmental organization (NGO) for scientific and technological professionals in China (China Association for Science and Technology 2018). CAST provides the link between the Communist Party of China, the Chinese government and China's science and technology community. It consists of 210 member societies spread across the country providing a network of scientists, engineers and other science and technology professionals.

The history of CAST can be traced back to the founding of the People's Republic of China in 1949. Thereafter in 1950, two new national organisations—the All-China Federation of Natural Science Societies and the All-China Association for Science Popularisation were formed. Subsequently in September 1958, both bodies decided at a joint congress to amalgamate and become one organisation: The China Association for Science and Technology (CAST).

CAST actively works to popularise science and technology amongst educators, learners and the public at large. Furthermore, CAST is also actively involved in nurturing talent amongst the science and technology community.

The Bridging Science Museum with Schools programme, implemented by the Children and Youth Science Centre, is in clear alignment with several identified major tasks of CAST, including:

- to carry forward the scientific spirit, popularize scientific knowledge and disseminating scientific ideas and methods according to *the Law of the People's Republic of China on Science and Technology Popularization*; uphold the dignity of science, promote the application of advanced technologies, encourage and organize science educational activities among children and youth, and improve the scientific literacy of all citizens.
- to carry out academic exchanges, activate academic thinking, promote the development of all scientific disciplines, and stimulate independent innovation.
- to carry out continuing education and training programs.

Initiated by CAST and the Ministry of Education in 2006, the Bridging Science Museum with Schools programme aims to build up a national network of Science, Technology and Engineering (STE) museums and centres to promote informal science education. The initiative addresses the needs of science education reforms by increasing the capacity of science centres and museums to develop educational programmes for school students, teachers and the public.

The Bridging Science Museum with Schools programme was organised by CAST and the Ministry of Education in 2006, targeting science museums, history museums, youth activity centres, primary schools, middle schools and high schools throughout the country. More than 90 STE museums and centres participated in the pilot programme. The programme aims to urge science museums to develop and offer science education actively, to meet the needs of science education in primary schools and high schools, and act as a catalyst to meet the need of science education reform of the schools. Specific goals of the programme include:

- Improve the visitor experience in science museums.
- Integrate various resources of the science museums in the science curriculum.
- Provide support to the science classes, field activities and research studies of schools.
- Provide training for the science education trainers.
- Improve the quality of extra-curricular science education.
- Provide training for professional science education teachers.
- Organise exhibitions and competitions of the education programmes of the science museums countrywide and provide an opportunity for science museums to exchange ideas.

To date 140 museums have been actively involved in this programme, covering all provinces, autonomous regions and municipalities in the country except for Tibet. This is impressive, as museums voluntarily participate in the Bridging Science Museum with Schools project. Each museum enjoys great flexibility on how to carry out the programme. Some programmes are deployed in museums and others

Fig. 5.1 Students preparing to enter the EScientia exhibit as part of the CAST: Bridging Science Museum with School programme

are organised in schools. A museum actively engaged with the local community can serve several schools in the city in which it is located. In addition, the train-the-trainer programme consists of face-to-face and on-line training and has had an impact on more than 300,000 trainees nationwide (Figs. 5.1 and 5.2).

5.3 IEEE's Informal Education Programme

IEEE has a long history of supporting and promoting engineering and engineering education among pre-university teachers, students, university students and the public. IEEE's involvement in informal education, through science and technology museums, provides the organisation with the opportunity to make an on impact students, teachers, parents and the public by exposure to and engagement with IEEE fields of interest.

A focus on informal education, planned and executed by the IEEE EAB, aligns with IEEE's overall Strategic Plan 2015–2020 in the following ways (Institute of Electrical and Electronics Engineers, Inc. 2018):

Fig. 5.2 Train-the-trainer programme as part of CAST: Bridging Science Museum with School programme

- Core Values:
  - Trust: being a trusted and unbiased source of technical information and forums for technical dialogue and collaboration.
  - Growth and nurturing: encouraging education as a fundamental activity of engineers, scientists, and technologists at all levels and at all times; ensuring a pipeline of students to preserve the profession.
  - Service to humanity: leveraging technology and engineering to benefit human welfare; promoting public awareness and understanding of the engineering profession.
  - Integrity in action: fostering a professional climate in which engineers and scientists continue to be respected for their exemplary ethical behaviour and volunteerism.

Key initiatives:

- Develop programmes in public service focused on knowledge and technology in IEEE's fields of interest related to public policy and humanitarian efforts.

Currently IEEE's presence in informal education is in ten science and technology museums in seven countries. IEEE exhibits in these locations are having an impact on more than 550,000 students, teachers and the public each year.

Through science and technology museums IEEE has an opportunity to take the lead in communicating and demonstrating how science, technology, engineering and mathematics work and are affecting society around the world.

A clear opportunity exists for IEEE, and other engineering societies, to engage and inform the public about STEM and associated careers through science and technology museum environments.

5.3.1 IEEE/EAB's Work to Date in Informal Education

IEEE's work in informal education, planned, developed and delivered according to the IEEE EAB, is guided by the following objectives:

- Create impact in informal education spaces by developing an assortment of hands-on science and technology centre exhibits and heightening the visibility of IEEE.
- Encourage interest in designing, innovation, and computing and related careers among pre-university students, their teachers and the public.
- Contribute to public awareness and understanding of electrical and computing technologies and their applications.

IEEE's EAB has developed and piloted two informal education programmes: IEEE E-Scientia and IEEE Low-Cost Exhibits.

- IEEE E-Scientia is a standalone museum exhibit for pre-university students aged 12–16, which introduces engineering through a guided simulated space flight.
- IEEE low-cost exhibits are hands-on self-guided exhibits targeting students aged eight to 18, with the objective of illustrating the fundamentals of science, technology, engineering and mathematics. More than 150 IEEE members, student members, high school students and professional staff have created 17 low-cost exhibits. These exhibits have been reviewed and evaluated by multiple experts in the field of informal education. Exhibits developed to date include:
  - Virtual physics laboratory
  - Demonstration of eddy currents
  - Boolean adder
  - Industrial robot
  - Force on current-carrying conductor (Fleming's Left-hand Rule)
  - Electrical resonance
  - Electromagnetic induction (linear variable differential transformer)
  - Life sciences
  - First millimetre wave experiment
  - Text-to-speech conversion
  - Computer-based language learning
  - Biometric and DNA identification demonstration
  - Green energy
  - Total internal reflections in prisms

– Fibre-optic communications
– Nanotechnology
– Discovery of the Raman effect.

IEEE E-Scientia and the IEEE low-cost exhibits programmes are experienced by more than 550,000 students, educators and the public annually. IEEE has cooperative agreements to field low-cost exhibits and/or E-Scientia in seven countries, including:

- Sci-Enza at the University of Pretoria, South Africa: E-Scientia and low-cost exhibits
- B.M. Birla Science Centre in Hyderabad, India: E-Scientia and low-cost exhibits
- Espacio Ciencia in Montevideo, Uruguay: E-Scientia and low-cost exhibits
- Shanghai International Sci-Tech Exchange Center, China: E-Scientia
- Universidad Nacional Autónoma de México (UNAM), Mexico City: E-Scientia
- Nehru Science Centre in Mumbai, India: Low-cost exhibits
- National Science Centre and Museum in Delhi, India: Low-cost exhibits
- National Council for Science and Technology, Nairobi, Kenya: Low-cost exhibits
- Centre for Science in Society, Cochin University of Science and Technology at Kochi, Kerala, India: E-Scientia
- National Institute of Higher Education, Research, Science and Technology in Trinidad and Tobago: E-Scientia.

There are numerous benefits for the end-users of these programmes, IEEE members and IEEE. These include: (1) the ability for pre-university students to learn about science, technology, engineering and mathematics in a fun, hands-on informal learning environment; (2) opportunities for pre-university educators, parents and the public to engage with students about science, technology, engineering and mathematics as programmes of study and career choices and to supplement the teaching and learning in STEM curricula; (3) opportunities for IEEE members to get involved in outreach at their local science and technology museum; and (4) an increase in the overall IEEE organisation benefits through increased visibility in the local community brought about by public informal education learning environments.

5.3.2 *Working Toward the Future*

IEEE has been involved with informal education for several years. This involvement generally entailed working with local IEEE members and their communities' science and technology museum personnel. What it did not include was a broader and deeper view of what IEEE could/should focus on to expand its footprint in the informal education area.

In late 2015 IEEE EAB organised and hosted an IEEE Exhibits Programme Strategic Summit focusing on:

> (1) creating global impact within informal education spaces by developing a variety of hands-on science center exhibits and to heighten the visibility of the IEEE and

(2) encouraging interest in science, technology, engineering and mathematics and associated careers among pre-university students, teachers and the public.

This two-day informal education strategic summit brought together 31 attendees including IEEE members, educators, and professionals, from ten science centres and one association to discuss future strategies for the IEEE's work in informal education. Organisations represented include:

- Sci-Enza, South Africa
- Espacio Ciencia, Uruguay
- Singapore Science Centre
- The National Institute of Higher Education, Research, Science and Technology - Trinidad and Tobago
- Kenyatta University Center, Kenya
- Centre for Science in Society, Kerala, India
- Boston Museum of Science, MA, US
- B.M. Birla Science Centre, Hyderabad, India
- Baltimore Electronics Museum, MD, US
- Discovery Cube of Los Angeles, CA, US
- The Southern African Association of Science and Technology Centres.

Summit attendees agreed on the following guiding principles when working in the informal education area:

- Provide an educational and interactive experience for pre-university students, their teachers, parents and the public.
- Illustrate fundamentals of science, technology, computing, engineering and mathematics and include their applications.
- Develop exhibits that are:
  - Simple
  - Well-documented and easy to duplicate
  - Modular
  - Inclusive of plans and drawings
  - Sturdy
  - Clear, visible and offering engaging explanations
  - Inexpensive
  - Versatile (Figs. 5.3, 5.4, 5.5, 5.6 and 5.7).

5.3.3 *Informal STEM Education: What the Future Could Be*

The summary and outline below provide all engineering societies with a roadmap to expand or establish a presence in the informal education area.

Fig. 5.3 IEEE EScientia exhibit at the Espacio Ciencia in Montevedeo, Uruguay

Fig. 5.4 Inside the IEEE EScientia exhibit at the B.M. Birla Science Museum in Hyderabad, India

Fig. 5.5 IEEE. EScientia at Sci-Enza, Pretoria South Africa

Fig. 5.6 IEEE Low-cost Exhibit at the B.M. Birla Science Museum, Hyderabad, India

Fig. 5.7 Inside the IEEE EScientia Exhibit at the Espacio Ciencia, Montevideo, Uruguay

Science centre professionals who attended the 2015 summit strongly recommended that engineering societies take the lead in expanding their efforts in informal education leveraging science and technology museums.

The overall focus of the future is a significant increase in the awareness of parents, students, teachers and the public about IEEE STEM fields of interest to encourage and influence students to pursue STEM as an academic programme of study and career choice and to ensure a pipeline of young professionals in STEM disciplines. The overall focus of engineering societies' future in informal education should include:

- A description of engineering, engineers and the contribution of engineers to society; and
- Resources for those who are or might consider STEM as a career path, or those looking to support those considering a STEM career path, e.g. teachers, parents, university students and pre-university students aged eight to 18.

The outcomes of the IEEE's informal education programme should be to:

- Increase awareness of engineering, with the emphasis on engineering, computing and technology (ECT), with younger students (ages eight and younger)
- Effect a change in knowledge.
- Focus on providing resources for those considering a career in engineering.
- Provide a learning experience for students, ages eight to 18, to learn about what different types of engineers do; to think more positively about engineering, breaking through stereotypes; to feel capable of doing engineering; and to experience engineering (problem-solving) through interactive delivery options.

- Provide information to help students decide about academic programmes of study and career paths and discover talents.
- Help students learn basic scientific engineering principles.

The delivery strategies of IEEE's work in informal education should include:

- A facilitated programme;
- An un-facilitated experience; and
- Physical interaction with engineering.

A proposed set of steps to expand IEEE's work in informal education is outlined below. This work is viewed as a five-year effort. The activities will focus on five areas:

(1) Assess the "market" to determine the most effective geographic deployment; to determine appropriate delivery venues, including science and technology museums, community centres, libraries and more; and to determine the most effective activities.
(2) Obtain the engagement and agreement of IEEE members and the appropriate delivery venues.
(3) Enhance the content collection.
(4) Effect formal and informal education integration.
(5) Create on-line activities, in conjunction with TryEngineering.org, with the emphasis on ECT.

The following outlines the timelines and activities for each focus area:

1. Assess the "market"

   - Year One: Engage with an appropriate research firm to assist in understanding the market landscape and in determining: (a) the most effective geographic strategy; (b) the most effective delivery venues; and (c) the types of activities to develop.
   - This work should be completed within the first six to eight months after the start of the project.

2. Obtain the engagement and agreement of IEEE members and the appropriate delivery venues

   - Year One: Establish an advisory group comprising active and tenacious IEEE members and representative(s) from science centres, community centres, etc., to support the mission of the programme and advise on programme activities.
   - Ongoing: Identify and seek funding to support programme activities.
   - Years Two to Five: Promote the programme to local IEEE organisational units to encourage and support partnerships with appropriate venues at the local level.
     - Customise the agreements based on the inputs of the appropriate venues.

3. Enhance the Collection

   - Years One and Two: Evaluate and enhance the collection website.
   - Years Three and Four: Productise content, market "right to use" licenses.
   - Conduct market research to determine if there is a market to generate revenue from exhibits products.
   - Year Five and beyond: Reinvest to create new exhibits to expand collection.
     – Working with experts in exhibit evaluation, evaluate exhibits in the collection and make enhancements, if necessary.
     – Develop a single online resource of information about the exhibits (i.e. website).
     – Create complementary virtual exhibits.
     – Incorporate mechanisms for continuous data collection to drive ongoing evaluations.

4. Formal and Informal Education Integration

   - Years One and Two: Develop resources for educators, lesson plans, field trip guides.
   - Years Three and Four: Conduct in-person workshops at science centres.
   - Year Five and beyond: Conduct in-person workshops, create virtual workshops.
     – Develop resources guides for teachers (e.g. teacher manuals, lesson plans, field trip); follow TryEngineering.org/TryComputing.org procedures to vet content.
     – Create professional development workshops (in-person and virtual) to be hosted at appropriate venues for local STEM teachers.

5. Create on-line activities, in conjunction with TryEngineering.org, with the emphasis on ECT.

   - Year Four and beyond: Establish partnerships to create on-line activities that demonstrate concepts in ECT.
     – Include these activities on TryEngineering.org to add to the portals 1.5 million annual visitors.

Science centre professionals unanimously recommended that IEEE take the lead in expanding informal education opportunities in science, technology, engineering and mathematics globally.

IEEE can and should play a key role in engineering, computing and technology education through informal education strategies. IEEE can continue to engage in this work by leveraging volunteers, conducting a "market" assessment and partnering with museums and professional organisations working in informal education. This work will result in a variety of venues that can be leveraged, as well as a variety of activities and programmes to influence teachers, students, parents and the general public in the most effective way.

5.4 Chapter Summary

Science and technology museums are an effective means of introducing STEM subjects to students and to the adults who influence them. In supporting appropriate informal education venues, engineering societies can raise awareness of the engineering profession and enhance its presence in local communities. This will demonstrate the value of the profession to society and be of direct benefit to the communities in which such programmes have been launched. Members of engineering societies have an opportunity to volunteer and engage the community in an informal education setting. Working in the informal education area gives the local organisational units of engineering societies the opportunity to demonstrate leadership in fostering society-industry-science museum cooperation.

References

China Association for Science and Technology (2018) http://english.cast.org.cn/. Accessed on 10 July 2018

Institute of Electrical and Electronics Engineers, Inc. (2018) https://www.ieee.org. Accessed on 16 July 2018

Koivu T, Myllykoski M (2015) Idea share—a white paper on the changing role of science centres. Spokes #14. European Collaborative for Science, Industry and Technology Exhibitions. Brussels, Belgium

National Research Council (2009) Learning science in informal environments: people, places, and pursuits. The National Academies Press, Washington, DC

National Research Council (2015) Identifying and supporting productive STEM programmes in out-of-school settings. The National Academies Press, Washington, DC

Chapter 6
Considerations for Engineering Societies

Abstract Engineering societies have developed programmes and activities designed to have an impact on engineering education. A wide range of possible approaches exist for engineering societies to consider in their quest to contribute to engineering education. This chapter describes several factors for the consideration of engineering societies that are planning to expand and/or establish a portfolio of programmes and activities to promote and support engineering and engineering education through social innovation.

This publication has described many programmes and activities that have been developed and deployed by engineering societies to promote and support engineering and engineering education through social innovation. A few examples were BCB supported by SAIEE, Bridging Science Museum with Schools developed and implemented by CAST, EPICS in IEEE and A World in Motion, organised and carried out by SAE.

Engineering societies are to be commended for their work in this area and for meeting their corporate social responsibilities and developing a portfolio of effective and replicable resources for the benefit of society. This is reinforced in the 2017 National Academy of Engineering paper reviewing engineering societies' websites to identify engagements in engineering education. The paper's analysis "... found that engineering societies, in line with their core purpose, work to impact engineering education both indirectly and directly."

Engineering societies that are planning to expand and/or establish a portfolio of programmes to promote and support engineering and engineering education through social innovation have many factors to consider, including:

- Establishing clearly identified short-range, mid-range and long-range goals and outcomes;
- Developing programmes and activities that follow the principle of "think globally, act locally";
- Identifying appropriate funding mechanisms;
- Identifying and developing partnerships, both internally and externally;
- Obtaining "buy-in" from key stakeholders in the society, both members and professional staff;

D. Gorham and N. Nwulu, *Engineering Education through Social Innovation*, Lecture Notes in Networks and Systems 108, https://doi.org/10.1007/978-3-030-39006-8_6

- Identifying effective strategies to recruit and organise local members who are interested in working in this area;
- Analysing and identifying the professional staff's capacity, both knowledge/expertise and available time;
- Planning for a comprehensive web presence;
- Developing a plan to collect impact/success metrics for each programme and/or activity;
- Determining how to design programmes and activities that address what participants should know (theory) and be able to do (practical application); and
- Developing and/or expanding an internal infrastructure to include society members and professional staff.

The above list is not intended to be exhaustive, nor is it intended to outline a daunting undertaking. Engineering societies that are considering expanding and/or establishing a portfolio of programmes and activities that promote and support engineering and engineering education need the support and commitment of their membership and professional staff.

But where to start? Public awareness? Informal education? Pre-university education? Programme accreditation? Standards education? University education? Public policy?

The 2010 UNESCO report, "Engineering: Issues Challenges and Opportunities for Development", identifies four areas that can serve as a guide for engineering societies looking to expand and/or establish programmes and activities that support and promote engineering and engineering education through social innovation:

- Improve public awareness of engineers and engineering, which will overtime increase the perception of engineering and technology as major drivers of innovative ideas for societal development.
- Enhance information on technology and engineering which is needed to better obtain and analyse engineering indicators.
- Modify engineering curriculum and pedagogical approaches to further incorporate advanced problem solving techniques.
- Increasingly deploy engineering and technology to solve practical, real world problems enumerated in the sustainable development goals (SDG's).

Additional guidance is found in the RAE 2014 report, "Thinking like an Engineer: Implications for the Education System". This report identifies three core conclusions that suggest that how engineers think, and the implementation of engineering habits of mind may inform the development of curricula, programmes and activities for target audiences:

- The most important finding from this research is that teachers of engineering really engaged with the question: 'How do engineers think?' Our work has highlighted a core idea that engineers make 'things' that work or make 'things' work better. Our model of EHoM provides a fresh way of exploring the teaching of engineers.
- At various levels the engineering teaching and learning community—school, college and university—agrees that understanding better how engineers think

could help teachers of engineering when they are constructing curricula, selecting teaching and learning methods and assessing learner progress on a course.
- We also conclude that understanding the thought process of engineers will increase the attractiveness and effective presentation of engineering careers to the younger generation.

Whether one is a member of an engineering society, an engineering educator, a professional staff member of an engineering society or someone with a general interest in engineering and engineering education, we hope this publication has provided you with additional knowledge and insight into what exists and what is possible in promoting and supporting engineering and engineering education through social innovation.

References

National Academy of Engineering (2017) Review of the literature on engineering society involvement in undergraduate engineering education. The National Academies Press. Washington, D.C., USA, National Academy of Sciences

Royal Academy of Engineering (2014) Thinking like an engineer: implications for the education system. London, United Kingdom

United Nations Educational, Scientific and Cultural Organisation (2010) UNESCO report: engineering: issues challenges and opportunities for development. Paris, France